Fuzzy Sets Theory Preliminary

Hao-Ran Lin · Bing-Yuan Cao
Yun-zhang Liao

Fuzzy Sets Theory Preliminary

Can a Washing Machine Think?

Translated and Compiled by Jun Xu; English Proofreading
by Pei-hua Wang

China Science and Education Press

Hao-Ran Lin
Shanghai Chinese School
Shanghai
China

Bing-Yuan Cao
Foshan University
Foshan
China

and

Guangzhou University
Guangzhou
China

and

Guangzhou Vocational College of Science
and Technology
Guangzhou
China

Yun-zhang Liao
School of Mathematics and Information
Guangzhou University
Guangzhou
China

Jun Xu, Guangzhou Vocational College of Science and Technology, Guangzhou, China, harryxujun@163.com;
Pei-hua Wang, Guangzhou University, Guangzhou, China, phwang321@163.com

ISBN 978-3-319-88987-0   ISBN 978-3-319-70749-5   (eBook)
https://doi.org/10.1007/978-3-319-70749-5

Printed on acid-free paper

This Springer imprint is published by the registered company Springer International Publishing AG
part of Springer Nature
The registered company address is: Gewerbestrasse 11, 6330 Cham, Switzerland

Mr. Hao-Ran Lin is sitting together with Prof. L. A. Zadeh, founder of fuzzy mathematics from University of California.

The Fuzzy Information and Engineering Branch of China Operations Research Society was founded in August 2005 at the meeting held in Guangzhou University. During the meeting, Mr. Hao-Ran Lin gave his "Fuzzy Mathematics Preliminary (Further Curriculum for Senior Middle School Mathematics)," a senior middle school textbook, to Prof. L. A. Zadeh from University of California, to have a look at it. Prof. Zadeh was very happy and asked whether there was an English version.

12 years later, the English version concerned by Zadeh is published finally by Springer (German) Press.

Foreword I

Mr. Hao-Ran Lin has been teaching fuzzy mathematics at middle schools in Shanghai, China, for over 10 years, which has been recognized as an unusual and exciting event and praised especially by Prof. L. A. Zadeh. Mr. Lin is a man who is walking in front of time, and he has done a work ahead of time. Fuzzy math was not accepted by a part of mathematicians; so far, this subject has not been taught yet officially as a basic course in universities, but as early as in 1997 he started to spread the idea of fuzzy mathematics to middle school students. I respect him very much for his practice in advance and his seeking innovation. In front of waves in the age of big data, we need more teachers and scholars to walk in front of time and bring young people to the front of time.

Mr. Hao-Ran Lin, Prof. Bing-Yuan Cao, and Prof. Yun-zhang Liao co-write this book "Fuzzy Sets Theory Preliminary," which has the milestone significance on the popularization of fuzzy sets theory. This book has the following outstanding features: (1) combining vividness, friendness, and intimateness; (2) being intuitive, simple, and clear; (3) easy for reading, understanding, and exercising. The three features reflect Mr. Lin's teaching spirit, while Mr. Lin is not only a prominent teacher, but an excellent scholar also; this book has another most important feature: (4) this book comprehensively and accurately reveals the essence of fuzzy set in the development of artificial intelligence.

This book is a special treasure for middle school students; it is not only for young students, but for readers of all ages and professional ones also. I would like to recommend this book to all friends.

Fuxin, China
Canyon, USA
March 2017

Prof. Pei-zhuang Wang
Liaoning Technical University;
West Texas A & M University

Foreword II

Professor L. A. Zadeh, a famous cybernetics expert at University of California, first proposed the concept of fuzzy sets in 1965, laying the foundation for fuzzy theory. This theory has been paid rapid widespread attention to because it has the advantages of simpleness and powerfulness in dealing with complex systems, especially human intervention systems. For decades, this field, from theory to application, from soft to hard technologies, has achieved fruitful results while it has been suffering uninterruptedly from authoritative doubtful voices.

The traditional middle school mathematics has been trying to find an only correct key, while the fuzzy mathematics is trying to find the most satisfactory solution, which is not necessarily unique, but more adapted to the requirements of today's information age. In fact, the fuzzy theory is a product of the computer age. It is an invention and wide use of this very sophisticated machine that makes people a deeper understanding of the precision of limitation, promoting people to its opposite—the research of ambiguity. The invention of electronic computers is the greatest scientific achievement in the twentieth century. It is only in a few years that the impact has spread to every corner of the world and penetrated into human life in all important areas. The human brain plus the computer has a big step forward into today's information society. However, so far, the most advanced computer has also a fundamental flaw that it does not have the human brain's unique fuzzy inference and fuzzy decision-making ability, so that it cannot process the knowledge expressed by natural languages in fuzzy environments like human brain, and so that it cannot flexibly draw an approximate inference like human brain. That is to say, the computer still has only part of the intelligence of human brain. In order to solve these problems to make the computer smart like human brain, the computer is required to have the ability to process fuzzy information, which requires people to have a better understanding of fuzzy concept and fuzzy inference. Professor Zadeh has made a creative research for this. He gives us a brand-new calculating concept with the development of words and soft calculation, which opens a door to imitate and understand the thinking and inference of human brain.

A late start but rapid development appears to China's research in fuzzy logic theory and application. The earliest research began in 1976. In 1981, we set up

"The Fuzzy Mathematics and Fuzzy System Branch of China's System Engineering Society" and founded the *Fuzzy Mathematics*, the world's second academic journal in fuzzy mathematics. In 1987, it was changed into *Applied Mathematics*. In the same year, the society founded another journal *Fuzzy Systems and Mathematics* again.

Since 1997, Mr. Hao-Ran Lin has instructed the enlightenment teaching of "Fuzzy Mathematics Preliminary" at middle schools in Shanghai, China, for 11 years, which has repeatedly been praised by experts and scholars at home and abroad and welcomed by the students. The teaching is often reported and concerned by the media and the professional journals. It has won many awards in the innovation competitions at all levels for many times. Fuzzy mathematics involves a wide range of fields, of which fuzzy control is one branch. It is a simple and effective intelligent control. It is maneuverable and practical to middle school students. Therefore, the course uses the fuzzy control as a breakthrough to organize teaching. In addition, the artificial intelligence-controlled robots can do a lot of things that people can not do. In the early year, Mr. Lin led students to participate in the Robot Football Match for World Cup. Now, the robots have made greater progress than before with judgment, agility, and ball-controlling skills. The international sponsors announced confidently: In 50 years, the robot champion team of the Robot Football Match for World Cup would compete against a team made up of human players! Who would win? Would intelligent robots reach or even exceed human brain? Suspense and curiosity have greatly inspired the students' interests and hearts … at the end of this course, many students wrote at feedback: "I see the future, it is running!"

Fuzzy mathematics teaching is difficult to promote here although its enlightenment and popularization have been welcomed by the students for 11 years. It is due to the reasons known as an existing college entrance examination system in China. In order to promote the effective popularity of fuzzy mathematics in middle schools, *Fuzzy Sets Theory Preliminary—Can a Washing Machine Think?*, a further-studied textbook for senior middle school mathematics, is written by Hao-Ran Lin, Bing-Yuan Cao, and Yun-zhang Liao, which is examined and published jointly by the world-renowned German publishing house: Springer and China Science and Education Publishing House in Chinese and English versions. It is the world's first fuzzy mathematics enlightenment textbooks in the world for middle school students. We eagerly expect middle school students to be free from the examination-oriented education, eager to explore, to foster themselves in their innovative spirits, be eager to explore, and become national innovative talents!

Changsha, China   Hao Wang
Mid-Summer 2013   National Defense University
                                                                              of Science and Technology

Brief Introduction

The Chinese book "*Fuzzy Mathematics Preliminary*" has been used at the middle schools in Shanghai, China, for more than 10 years. Now it is translated into English. According to the comments of Springer's experts, "This is quite an interesting project as there are no such basic texts" as, "for instance, fuzzy topology, fuzzy category theory, etc., at a higher level". So Springer's experts suggested to change the book's name into "Fuzzy Sets Theory Preliminary". The book has carefully selected contents in order to achieve the roles of enlightenment and popularization. It consists of five chapters. Chapter 1: Human Brain, Computer and Fuzzy Mathematics; Chapter 2: Matrix, Fuzzy Relation, and Fuzzy Matrix; Chapter 3: Fuzzy Control; Chapter 4: Fuzzy Statistics and Fuzzy Probability; and Chapter 5: Fuzzy Linear Programming. At the same time, it has some concise, interesting, and profound reading and thinking materials, and a certain amount of exercises so as to make it an informative and interesting textbook. This book can be used not only as a further-studied mathematics textbook in senior middle schools, and in vocational colleges, but also as a primer for learning fuzzy mathematics.

Contents

About the Authors

Hao-Ran Lin graduated from the Department of Mathematics, Shanghai, Normal University in 1968 and has long been engaged teaching in middle school mathematics. Since 1997, by using teaching materials, he tried fuzzy mathematics enlightenment and popularization work first in ordinary middle schools and published twenty papers in the "Mathematical Bulletin," "People's Education," "Shanghai Education," and other core journals in mathematics education in China. He propagandized enlightenment to secondary school students by popularization of fuzzy mathematics necessary and possible. In 2003, he was named the top ten annual figures in Shanghai education. In 2005, he asked Professor of L. A. Zadeh, the founder of fuzzy mathematics from University of California, whether the English version of this textbook could be published, which has formed this book today.

Bing-Yuan Cao was Professor and Doctoral and Postdoctoral supervisor in School of Mathematics and Information Science, Guangzhou University; Dean and Professor, School of Finance and Economics, Guangzhou Vocational College of Science and Technology, China. Second-level Chair RE Professor of Lingnan, Foshan University. He was the Former President of Fuzzy Information and Engineering Society in Operations Research Society of China, present Operations Research Society of Guangdong Province, and Operations Research Society of Guangdong, Hong Kong, and Macao. He was the President of China science and education press. He developed fuzzy geometric programming, published 7 books and 10 proceedings in Springer. He has published more than 180 papers (SCI index 40). He serves as the editor in chief of the 3 international magazines, and 5 deputy editors or member of the editorial board of magazine. He has won 15 Honors and Awards and has been in charge of 12 scientific research projects, training more than 30 doctoral, postdoctoral, and master's degree students home and aboard.

Yun-zhang Liao is a Professor and Doctoral Supervisor at School of Mathematics and Information Science, Guangzhou University, Guangzhou, China. His research interest is related to mathematics curriculum and teaching theory. In particular, he is interested in the cognitive psychology of mathematical application problems solving, international mathematics curriculum comparison, mathematical assessment, mathematics teacher education, and other fields. His work has been published in various research journals; more than 80 papers were published, including *Curriculum teaching materials, teaching methods, Journal of mathematics education*. He has presided over multiple national, provincial, and other kinds of research projects. In 2014, *the middle school mathematics teaching theory and practice of problem driving* was awarded the second prize of national teaching achievements.

Introduction

Sets are often arranged in the first chapter of senior middle school mathematics. Sets are original concepts in mathematics, which cannot be defined with other more basic concepts. So the sets are also known as the concepts without definition. They can only be defined by descriptive explanation. The objects in the sets are called elements, which can be anything. A set must have three characteristics: certainty, mutual difference, and disorder.

Defects in Sets

For example, which of the following objects can make sets?

   (i)   Pretty fabric;
  (ii)   All quite small numbers;
 (iii)   The boys who are taller than, or as tall as 165 cm in Class (1), Senior One;
  (iv)   The famous film stars;
   (v)   The boys who are taller in Class (1), Senior One;
  (vi)   All the numbers that is greater than 5.

   The answers are that only (iii), (vi) can constitute the sets with clear boundaries, instead of the others. Because these criteria, such as "pretty," "quite small," "famous," "taller," are very vague, they do not have the certainty, leading to uncertain "sets" with blurred boundaries.

Sets Expanded to Fuzzy Sets

In fact, uncertain objects can also be converted into certain objects under certain conditions. For example, in Class (1) Senior One, there are totally 25 boys, of whom 6 people are 155–160 cm tall, 10 people 161–165 cm, 5 people 166–170 cm, and 4 people taller than 171 cm, respectively. If we set the criteria as "taller" than 161, 166, and 171 cm, then the corresponding numbers of the sets are 19, 9, 4 people, respectively. In the same way, although the famous film stars are famous at different levels, they can also be separated by years or by different awards to ensure the number of people. These boundary-blurred, uncertain sets are called fuzzy sets in fuzzy mathematics. The certain condition—such as taller than 161 cm—in the fuzzy sets is called the threshold, denoted as λ. This new set cut down according to the condition is called cut set λ. This kind of cut sets can satisfy certainty, mutual difference, and disorder.

Enhancing Intelligence by Mathematics Development

So, why can a set be formed only by certain elements? Why can it not be extended to contain uncertainty, either? The uncertain objects are much more than the certain objects in the world. Would mathematics be too "incompetent" if it could only solve a few certain problems but leave out a lot of uncertain problems that must be resolved? As we entered into the information age, the computer was getting more and more incompetent to treat uncertain events with fuzzy boundaries. Many scholars were thinking about it with confusion. In 1965, Prof. L. A. Zadeh at University of California, USA, who has been engaged in automatic control, wrote the first paper on fuzzy sets. He has extended from classical boundary-certain sets to boundary-uncertain fuzzy sets and expanded from either of the two values {0, 1} of the classical boundary-certain sets into the infinite values [0, 1] of the boundary-uncertain fuzzy set. He has opened a new era of fuzzy mathematics. This is a milestone! But for the first decades, he has been questioned and criticized by some famous experts. Just as $x^2 \neq -1$, the imaginary number was suspected once by many greatest mathematicians, such as Descartes, Leibniz, and Euler. However, it is two mathematics amateurs: Wessel, a Norwegian cartographer, and Ar-gand, a French accountant, who have rehabilitated the imaginary number. The two small potatoes have established a complex plane that opened a new world of modern physics. Fuzzy mathematics has gone through such a difficult road. After the 1990s, the application of fuzzy control has been popularized increasingly. Fuzzy mathematics is also accepted by more and more people. Because the set theory is the foundation of modern mathematics, many important branches of mathematics are based on the set theory. So once the sets are extended, a large number of new disciplines are also produced, such as fuzzy probability, fuzzy linear programming, fuzzy topology, fuzzy decision, and fuzzy cluster analysis. What makes people happier is that the computer's ability of imitating human brain has also become increasingly stronger and stronger. The intelligence of computers has made substantial progress. The artificial intelligence theory and technology have re-emerged from the bottom, showing an attractive prospect.

What is Intelligence? What is Artificial Intelligence?

Intelligence refers to the common ability to learn and use automatically the knowledge learned to solve problems effectively. The humans' ability performed in daily activities can all reflect the intelligence, such as language skills. Put a child in

Berlin, New York, or Beijing, he will speak German, English, or Chinese fluently. He neither requires the adults' intervention, nor modifies his "program." The key is that children are born with brain self-learning functions. This innate intelligence enables them to adapt to their environment, and learn and use automatically the local language soon.

Artificial Intelligence, abbreviated as AI. It is a new technological science to research and develop the theories, methods, techniques, and application systems used in simulating, extending, and expanding human intelligence. Artificial intelligence is a branch of computer science that makes persons try to understand the essence of intelligence and produce a kind of new intelligent machines that can react in a similar manner of human intelligence. Artificial Intelligence tries to simulate human awareness and thinking information process, while fuzzy mathematics is one of the best tools to simulate the thinking of human brain.

"Can a Machine Think?"

These are opening words of an essay in a monthly journal *Thinking* published in 1950. At that time, the article entitled "Computer and Intelligence" aroused a great controversy. The author is Alan Mathison Turing, a British mathematician known as "Father of computers."

Since then, in the past 50 years, the computer has undergone enormous changes. It can not only calculate, store, and remember information, but also be faster and more accurate in calculation and longer and more in remembrance than human beings. It can simulate many human productive and social activities. It is so smart that people would regard this calculative machine as the computer—the brain made up of electronic components! Moreover, if something happens, our first action is to go to a computer and look it up with the computer. Today with Internet, the computer has all become our daily lives in leisure, entertainment, shopping, and working. People have been looking forward to the computer's success very much. However, by far the most advanced computer still has a fundamental flaw that it could not have the human brain's unique ability, such as fuzzy inference and fuzzy decision-making. It could not use natural language to express knowledge in a fuzzy environment. It could not draw an approximate inference as flexible as people do, nor could it use natural language to talk with people. In sum with a word, the computer still has only part of the human brain's intelligence. To solve these problems, we require that a computer would be able to handle fuzzy information so that the computer could be smarter like its owner. It requires people first to understand further the fuzzy concept and fuzzy inference. This course tries to open a new perspective for middle school students beyond traditional mathematics.

We start with a washing machine, for example, with analysis step-by-step about fuzzy concept, fuzzy inference and fuzzy control, fuzzy random, and fuzzy optimization preliminary.

Can a Washing Machine Think?

No, it can't. A general automatic washing machine cannot think. It cannot judge how many clothes to wash? It cannot judge all by itself whether it is going to wash the clothes made up of thin light chemical fiber cloth or of full cotton cloth? It cannot judge all by itself whether the clothes are made up of generally woven cotton or of heavy cotton like denim? It cannot judge all by itself whether they are daily-washed clean clothes, or dirtier or very dirty clothes worn for a long time? It cannot judge all by itself whether the stains are sediment types that are easy to clean or the grease types that are difficult to clean? So ignorantly, of course, it cannot decide all by itself how to wash these clothes. All of these should be done by its owner. After the owner presses the corresponding buttons, the automatic washing machine works according to the programs that are preset from pouring water and adding detergent automatically until the cleaning is finished. In fact, when an experienced housewife simply glances over the clothes, she could know by experience how much detergent this pile of clothes would require, and whether they would be washed with gently or hard scrub. Yes, there are many fuzzy concepts like these without clear boundaries in daily life everywhere, such as more or less (a lot, more, not many or much, not few or little, less, few or little), light or heavy (very light, lighter, not light, not heavy, heavier, very heavy), clean or dirty (very clean, cleaner, not clean, not dirty, more dirty, very dirty). Human brain can easily identify these fuzzy concepts and easily make a judgment, but the computer cannot do these although it is good at accurate calculation. Can we teach the computer to judge all by itself fuzzy concepts, such as a lot, more, and less; dirty, very dirty, more dirty without too much difficult calculation, so that it could make decisions all by itself on how to clean? The first three chapters of the textbook will explain these in detail.

Development and Prospects of Artificial Intelligence

Since the late 1980s, owing to the application of a variety of smart technology including fuzzy technology, language and word recognition have developed to reach a practical level. It is not news that the computer has been able to listen and speak, but it is still an important problem that must be resolved in the man-machine communications: The computer not only should be able to listen and speak in form, but also it needs to understand the true natural languages. Therefore, if people do not understand the mechanism with which the human beings process natural languages, it is of course difficult to "teach" computers to understand and process the natural languages. It is the fuzzy theory's contribution to linguistic research that membership functions are used to describe the semantics of natural languages. In addition, the computer can easily have a memory of a thick book with millions of words and repeat them without a word mistake, but it is very difficult for a computer to write an abstract of 500 words based on the original book because the ability to generalize is the human intelligent activity at a higher level. To solve this problem, different scholars from different perspectives have put forward a number of different methods. The fuzzy logic and the artificial neural networks belong to the intelligent technology with promising prospects. The fuzzy logic is essentially the variable structures and the nonlinear controls with strong knowledge expressing ability, while the artificial neural networks have good self-learning functions.

In the late 1950s appeared a "fantastic" attempt to match the computer science with the evolution theory, which has attracted a scholar. As a great computer scientist, he profoundly understood: "A living organism is an expert to solve the problems. What the organism has performed is enough to make the best computer programs ashamed. The phenomena have especially made computer scientists feel embarrassed. It would take computer scientists several months or even several years to work mentally on some algorithm while the organism can acquire this ability through the obvious non-orientated mechanism of evolution and natural selection." He is Prof. J. H. Holland at University of Michigan. After years of hard efforts, he with his group of students proposed and developed an algorithm—genetic algorithm, also known as genic algorithm, which is the optimization program simulating the natural selection and the genetic mechanism. In the early 1990s, the fuzzy neural networks have been introduced into genetic algorithm. Their combination forms a new hybrid intelligent system. Many scholars have pointed out: The future hybrid intelligent system should be based on soft calculation (fuzzy logic, artificial neural networks, and genetic algorithm), rather than relying mainly on traditional value calculation that is hard calculation or accurate calculation.

Why do human beings have intelligence? How the human brain works? These are the questions that scientists and ordinary people are very concerned about. Although the science in the twentieth century has made significant progress and the neuroscience circles exploring the nature of neural activities have reached a single cell and molecular level, they are still poorly understood, quite shallowly. The intelligence that the human brain has is the crystal for which it has, after all, experienced hundreds of thousands of years of human evolution and millions of years of biological evolution. Although the artificial neural networks simulating the biological neural networks have made considerable progress, they still could not show the intelligence that people would expect. This narrates that the overall functions of cerebral neural systems are difficult to be explained by the mechanism of a single cell or an individual neuron. The integrated approach to study the brain functions has been on the agenda. It will be one of the most challenging tasks in the twenty-first century. To study the brain functions with integrated approaches, fuzzy theory will play an important role in it.

Can the Intelligence of Machines Excel the Man?

There is a Frankenstein attempt to do this. He is Prof. Hugo de Garis, an Australian, and has been called the "father of artificial brain," who has worked and studied in seven countries and has mastered four languages. He is now trying to learn Chinese. He came to China in 2006 and became a full-time professor at International Software College in Wuhan University. In 2007, his monograph *Smart History:*

Who Will Replace Humans as the Dominant Species was published by Qinghua University Press, in which he proposed to make artificial intelligence become several possible key technologies: Moore's law, reversible computing, nanotechnology, molecular-level engineering, artificial embryogenesis science, evolutionary engineering, and self-assembly. He was responsible for finishing two of the world four "artificial brain," one is in Japan and the other in Belgium, of which the intelligence of Japan's artificial brain was as smart as that of a domestic cat. He believed that it was only a matter of time and money that the artificial intelligence would excel human brain. It seems to be whimsical! (see Reading and Reflecting of this book)

For over half a century's brilliance, the computers' development in speed performance, reliability, calculation cost and volume reduction, etc., can be described by one number—"tens of billions times," which could be visualized by the fact that "more than twenty millions years has become one day at a second." This result has made many impossible things possible and come true. The extremely expensive cost has become negligible. Thinking those 500,000 years ago, our ancestors still stayed in Zhoukoudian cave and computers were born only fifty or sixty years ago; computers have made so earth-shaking changes. What will happen in the next 50 years? Thus, the international organizers confidently declared that "In the next 50 years, the robot champion team of the World Cup Robot Football Match will play a match with a team made up of men." It is really worth waiting to know who will win!

The future intelligent development needs the joint promotion of multi-disciplines and multi-directions. What we select here is just one of the parts to interest students in middle schools and enable them understand and accept the content. The computer's glory for half a century is supported on the basis of dozens of fundamental innovations (such as cathode ray tubes, integrated circuits, instructions as the processed data, Boolean algebra, relational databases, core memory, parallel algorithms, high-level language, software engineering, laser printers, and Internet). Unfortunately, none of them is invented by a Chinese. The inventor of core memory is Wang An, a Chinese, but a Chinese-American. We have had the four great inventions, such as Chinese compass, papermaking, printing technology, and gunpowder. Our great country has led the world for two thousand years. For the past nearly 100 years, we have fallen behind.

There are 20 major inventions in the twentieth century that have influenced the human life; they are shown as follows:

American inventions: washing machine, TV, rayon, tape recorders, electrostatic copier, computer, microwave oven, transistors, contraceptives, integrated circuits, robotics, and Internet.

British inventions: electric light, stainless steel, test-tube baby.

German invented the electron microscope.

Italian and Russian invented the wireless.

Belgian invented the plastic.

Japanese invented the monosodium glutamate and the liquid crystal.

The twenty-first century is an era of big data. The accurate mathematics based on traditional numerical computation and symbolic inference cannot analyze and deal with the big data rolled in from all aspects of information. And fuzzy theory has continued to grow up in the voices of doubt, while the AI will develop rapidly.

Chapter 1 of this book introduces how fuzzy math generates, comparing the computer with the human brain. A number of interdisciplinary knowledge will be involved, such as physical, chemical, and biological knowledge. And then, the traditional sets will be expanded to fuzzy sets. Some operations will be introduced. Chapter 2 introduces the whole fuzzy relation, fuzzy matrix, and its operations. Chapter 3 introduces the fuzzy control, taking a washing machine as an example. In this chapter, we hope students from middle schools to know how it undertakes to turn precise figures into fuzzy inference with the washing machine and thus understand how its intelligence works in the whole process. This chapter has been explained with more words. If the classes are not enough, teachers can choose some parts to teach in classes, leaving other parts for students for self-education. Chapter 4 presents preliminary fuzzy statistics and fuzzy probability and explains how it is applied in sentencing and other issues. In Chapter 5, the fuzzy linear programming follows the middle school textbooks. It mainly introduces the fuzzy linear programming graphical method and uses the computer with the Excel calculation, which inspects the validity of the model and methods by application of vegetarian issues. The Introduction and Chaps. 1–3 of the textbook are written by Hao-Ran Lin (Hengyi), a senior teacher in a middle school in Shanghai, China. Chapter 4 is written by Prof. Yun-zhang Liao at Guangzhou University, who is making teaching materials and methods research. Chapter 5 is written, and the whole book is compiled by Prof. Bing-Yuan Cao, Ph.D. (postdoctoral) Supervisor at Guangzhou University and Dean of Guangzhou Vocational College of Science and Technology. Prof. Pei-zhuang Wang, Ph.D. (postdoctoral) Supervisor at Beijing Normal University, who is a pioneer in the field of fuzzy mathematics in China, has taken time to review the textbook and given high praises, for which we appreciate deeply.

We sincerely hope that this course could make middle school students put their ingenuity into generating curiosity, interest, and innovation, rather than overwhelmed by examinations. And we also wish this course might show a hand to help new generations of China to stand up in the nations of the world!

Chapter 1
Human Brain, Computer and Fuzzy Mathematics

1.1 Fuzzy Mathematics

1.1.1 Old or Not Old

1. The science that analyzes and processes fuzzy phenomena and things with mathematical tools is called fuzzy mathematics.
2. What is the "fuzziness"?

   We have discussed the concept of "the elder" with students in the classroom.

Q: "I'm your teacher. I'm standing in front of you, am I old or not?"
A: Some students answer, "You are old", and the others, "You are not old".
Q: "What is your criterion to judge I am old or not?"
A: According to the "old face", "white hairs"….
Q: "Yes, forehead wrinkles, and skin laxity, can you use the number to describe this criterion quantitatively?"
A: The "ages", "older".
Q: "Do you think how old a person called an elder will be?"

Each student answers this question differently, some say, "60 or older", some say, "65 or older", others say, "A 70-year-old person is still 'a little brother' …."

For the same person, whether it is old or not has not a definitive answer. It differs from the traditional mathematics where there is only one correct answer, because "the elder" here is a concept of fuzziness.

1.1.2 How Many Grains Are in a Pile of Sand?

Another example is "a pile of sand." Can we call 1 or 2 grains of sand a pile of sand? (Certainly not)

© Springer International Publishing AG, part of Springer Nature 2018
H.-R. Lin et al., *Fuzzy Sets Theory Preliminary*,
https://doi.org/10.1007/978-3-319-70749-5_1

100, 200, 300 grains … (As the number is getting bigger and bigger, people begin to say yes, some people still say no.). Are 100 million grains "a pile"? (Every one says yes.)

Can we find a number big enough to make the boundary recognized as "a pile", such as 2,325,647, more than that something can be called "a pile", otherwise it cannot be called "a pile"? If so, 2,325,648 can make a pile, and 2,325,646 cannot be called a pile. Is it reasonable? The problem is that the concept of "a pile" is vague.

1.1.3   Fuzziness is Everywhere, Visible at Any Time

We can still find many similar vague things, such as: high, low; fat, skinny; large and small; many, few; beautiful, ugly; generous, stingy; cold, warm, hot; good, bad; cloudy, rainy, breezy; pink, milk yellow, light green …. All things that can be modified by an adjective are always called fuzzy things, because there is no clear quantitative limit described by an adjective. So does an adverb, such as: extremely —very—relatively—slightly—tinily … maximally, very hard, stronger, more shallowly, smiling, there is no clear quantitative limit modified.

Other examples for fuzziness are: serious, sloppy; close, far; warm, cold; happy, painful … which are concerned with people's feeling. Even if the same thing happens, everyone is feeling it differently with each subjective criterion. Each would draw a different conclusion. For cheapness and durability, it is concerned with everyone's economic ability. For the same commodity, some people think it cheaper and more beautiful while others would regard it more expensive; another example: gentleness or crankiness is related to everyone's personality and endurance. These are all fuzzy concepts.

Language also has ambiguity. This sentence "I want to eat candies" is clear in meaning, however, people also can understand fuzzy language, such as "I want candies", "want candies" or even "I, I, I, I want to eat candies", "candies, candies" etc. The human brain can understand that the last two sentences are spoken by people who stutter or by children. But when you "speak" to a computer, you cannot do so at random. Or it would refuse to accept it by "grammar error" as it could only accept typed statement in line with its rigid grammar, even if only a few words or punctuation were omitted. Because the early computers have been designed on the basis of traditional precise mathematics, the ambiguity of things and phenomena exists almost everywhere, every when visibly. Then, would people ignore this limitation of mathematics or strive to improve and develop mathematics?

1.2   Birth of Fuzzy Mathematics

Fuzzy mathematics is not imagined or inferred out by some brain-developed Frankenstein?

The process of civilization tells us as follows.

1.2.1 Mathematics Started from "Fuzzy"

When the early apeman climbed down from the tree, they only had hazy consciousness about the number, such as "many", "more", "much more" and etc. With improved tools and increased preys, they have to distribute and redistribute the preys. They counted originally from their own fingers to establishing a decimal number, which was a great leap forward. Looking back, "fuzzy" represents a backward productive force.

1.2.2 Economic Development and Mathematics Development

With development of tools and production development, mathematics also formed and developed systematically with the production development.

Thousands of years ago, the application of bronze and iron developed agriculture and seafaring, and formed geometry and trigonometry.

Hundreds of years ago, the application of the steam engine promoted the industrial revolution, generated logarithm to simplify the operations and produced functions, calculus….

The insurance industry appeared to improve the probability and statistics; the victory of the Allied Forces in the World War II, made the reputation of linear programming ….

With development of tools and productive forces, mathematics developed more and more precise and tight. But to the information age, after computers appeared, the precision pursuit has made mathematics more useless!

1.2.3 Traditional Mathematics Cannot Solve Fuzzy Phenomena

1.2.3.1 Bald Paradox

We can use common sets as a mathematical model to give a clear classification of things. However, for many things in the world, this classification is not applicable. Greeks have been aware of this for a long time. In everyday life, it is very easy with common sense to judge whether a person is a bald or not. But it would be difficult to judge a bald according to the number of a person's hairs under a strict definition of balding. Because it is meaningless that the difference with one or two hairs would determine whether a person is bald or not, it is unreasonable that a person with 500 hairs is judged as a bald while a person with 501 hairs is not. Therefore, there is no clear quantitative boundary to judge whether a person is a bald or not. Thus starts a bald paradox. We can use the modern logic language to express the bald paradox as follows:

Proposition A: A person who has one hair more than a bald is still a bald.
Proposition B: A person who has one hair less than a non-bald is still a non-bald.

In a common sense, Proposition A and Proposition B can be regarded as true propositions, but to start with these two propositions, we can draw the conclusion by mathematical induction that "all persons are bald" or "a bald is a non-bald".

Let n represent the number of a person's hairs, when $n = 0$, apparently the person is a bald. Suppose a person is a bald when $n = k$, then according to Proposition A, when $n = k + 1$, a person is a bald, too.

Deduced by it, when n is arbitrary integer, a person is a bald, that is to say, "all persons are bald".

This is an absurd bald paradox, but the mathematical induction is not wrong. Where is the problem? This paradox shows that a bald or a non-bald, the facile concept in the daily life could not be defined by the traditional precise mathematical language. The ancient Greeks left this pity down to us intactly, but what is worth pondering upon is that it is not only one case like this in life, which always has no clear boundaries. Many fuzzy cases exist everywhere and everywhen. We can also draw a similar paradox by the same precise mathematical induction that: a little rain has washed a bridge away; or a fast-going train has stopped at a station.... It would not be serious if only a few things like the bald case happened. But the problems are that the world, with which the modern science and technology are confronting, has become more and more complicated since we entered the information society. The fuzziness is always accompanied by complexity, which often makes people stare in perplexity. Could "versatile" Mathematics turn a blind eye into the fuzziness permanently?

1.2.4   Computer Limitation

1.2.4.1   Early Computer is Not as Smart as a Baby; Robot Turnkeys

Decades ago, the birth of a computer generated the third tidal wave. The computer —fast and precise, large memory capacity and robustness, is willing to execute the human instructions faithfully. It can do a lot of work that people can not do. People have had great expectations for it. Once something happens, the first thought is to open a computer to check it out. The computer would have the potential to replace the human brain, and it happens in all precise areas. But to fuzzy fields, it seems to be useless. For example, some scientists in foreign countries have developed a computer-controlled robot as a prison guard. The advantages of it: (1) it is very strong; (2) it cannot be killed; (3) it is of rigid integrity and devotion to duty; but it also has a fatal weakness, and sometimes it is hard to tell who the prisoner is and who the policeman is. If its program had such message as: "The prisoner has a heavy beard", as the "beard" is a vague concept, it would demand further information: (1) how long is the prisoner's beard? (2) How thick is the prisoner's beard? (3) How many beards are growing in a square centimeter on the prisoner's face that can be called heavy beards? And to how many digits is the number of beards accurate after the decimal point? It is known that the man's beards are growing every day, and the beards are changing every day. It was possible that the robot guard would let the prisoner go, and would lock the poor policeman up. So the intelligence of a robot is not as smart as a baby. A baby of a few months will be able to tell correctly who his or her mother is, and who is not. But the robot cannot.

When people tried to use a computer to process fuzzy information, they discovered that the computer could not recognize and understand this kind of natural language, such as "more", "very", "not too", which became the fundamental obstacles that the computer would imitate the human brain.

Many efforts have been made to improve the computer: (1) to increase the capacity; (2) to improve the processing speed. Today, these two targets have exceeded the human brain, but the computer is still not effective.

It could not work by improving the hardware, so people would have to improve the software accordingly, so that the computer could learn to distinguish things as the human brain could do. How to improve? Since the computer is a powerful tool for digital information processing, it can understand and accept digital information. Therefore, if we could turn the natural language information into the digital information that computer could accept, the problem could be solved. The fuzzy mathematics is created first to solve such a problem.

1.2.5 Structure and Thinking of Human Brain

In recent years, when a Finnish biologist dissected a human brain, he has found the basic elements—synapses that the human brain uses to transmit information. When the relevant information passed the synapses, they acted as switches. The difference between the synapse and the computer was that the synapse had the strong functions, which could not only open or close the nerve path, but also could partly open or partly close it according to the strength of information.

1.2.5.1 Human Brain; Nerve Cells

Figure 1.1 is a cross-sectional view of a human brain. The human brain is made up of billions of neurons. The function of a neuron (nerve cell) is much better than a "Pentium" computer. All the neurons have a diameter of about 10–50 μm of stubby body portions, called a cell body. The neuron has many projections, called "dendrites" besides the cell body, which differs from other cells of the body, and in which a long and thin part is called axon (Fig. 1.2). The task of the dendrite is to accept the signal. It is like a huge marina, where it receives day and night a lot of

Fig. 1.1 The brain is divided
into different parts

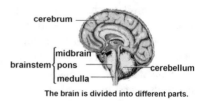

The brain is divided into different parts.

Fig. 1.2 The neurons

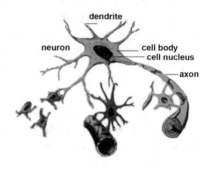

Fig. 1.3 Connection
diagrams of neurons and other
neurons

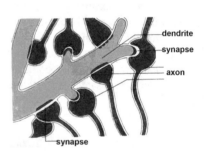

boats loaded with goods. The scattered signals are gathered together to the cell body, along with the dendrites. If the signal is strong enough, the cell body will produce a new electrical signal. The role of the axon is to transmit the electrical signal from the cell body to the next neuron, then to a remote "destination" (Fig. 1.3). Between dendrites and dendrites, and between dendrites and axons, they can synapse there. In the cerebral cortex, the pyramidal cells have more than 3000 synapses (Fig. 1.4).

1.2.5.2 Synapses—Powerful "Circuit" Switches

Small as a synapse is, it is composed of two neurons. The presynaptic membrane belongs to the upper level of neurons. Its function is to accept electrical signals, and then the electrical signals are converted into chemical signals. The postsynaptic membrane belongs to the lower level of neurons. Its function is that the recipient accepts chemical signals, and then the chemical signals are converted into the electrical signals.

Fig. 1.4 Synapses of neurons

synapse

Fig. 1.5 The schematic
diagram of the cell membrane

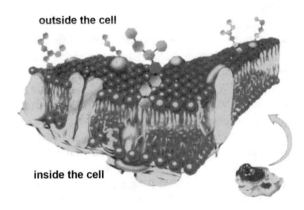

outside the cell

inside the cell

The brain cells work by the excitability of neurons and the transfer of electrical signals. The excitement is based on the bio-electrical changes.

The neuronal signal conduction in fact is that some protein molecules called "ion channel" in the cell membrane allow the ions to flow and produce electrical signal conduction (Fig. 1.5).

Where does the electricity in the cells come from?

Under general circumstances, the inner side of the cell membrane has negative charge distribution. Its outer side has an equal amount of positive charge distribution. In this case, the cell membrane appears like the "clear demarcation line" on both sides of which the positive charges and negative charges are confronted, "evenly matched." This is called polarization state. When the cell is in the quiet polarization state, there is on both sides of the cell membrane, the potential

Fig. 1.6 The resting
potential of the cell

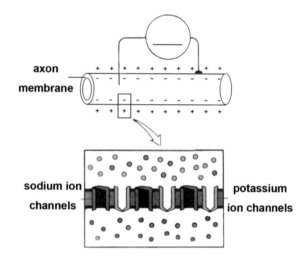

difference with the internal negative and the external positive charges, called the
"resting potential."

Under normal circumstances, the concentration of sodium ions and potassium
ions is different on the inner and outer membranes of neurons. The sodium ion
concentration outside the membrane is higher than that inside the membrane, but
the potassium ion concentration inside the membrane is higher than that outside the
membrane. When the cell is "quiet" (i.e. resting state), the membrane allows
potassium ions to pass through. Therefore, the potassium ion inside the membrane
can diffuse outward by the concentration difference inside and outside the cell
membrane while the sodium ion with negative charges inside the membrane can not
go out and in freely. In this way, when the potassium ions spread out to a certain
extent to balance the attraction of the inside negative charges, the diffusion stops.
The result is that it would form a potential difference between the two sides of the
membrane that is called the resting potential (Fig. 1.6).

When the cells are stimulated, they would make the cell membranes change the
ion permeability, making the ions transmembrane-moving and re-distributed on
both sides of the membranes. The potential change caused by the transmembrane
ion movement is the formation process of "action potentials". The forward conduction of the nerve action potential is called nerve impulses, also called nerve
discharge (Fig. 1.7).

Besides the sodium ion channels, the potassium ion channels, calcium ion
channels and chloride ion channels etc. have currently been found.

In general, each channel allows only one ion to pass through. In accordance with
the open-close characteristics of ion channels, the nerve impulse conduction in the
nerve fibers can be seen as the continuous "close-open-reclose" switching forms, by
which the process repeats the conduction forward from one nerve fiber to another
continuously.

Fig. 1.7 The action potential
of the cell

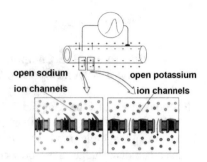

Thus, a neuron is a high-level computer. In the brain there are 10 billion computers, working with infinite amounts of information, resulting in the senses, and completing the action instructions to adapt the body to the environment perfectly. On the basis of these, the brain has all kinds of advanced features, including the transformation from the materials to the spirits.

From the above, we have described in details the cells of the human brain enlarge the functions progressively. Can you understand why the human brain's "switches" can be partially opened, or partially closed? Can you realize how the human brain can distinguish accurateness from fuzziness, but the computer cannot?

For example, because the electrical current only has two states: either passing or not passing, when the computer judges whether it is pink or not, it will open if it sees the pink (switching on), or it will close (switching off) if it sees the yellow or any other non-pink colors. It can only tell yes or no. It cannot distinguish different shades of pink so that it corresponds to the binary digits: on—1, off—0; the computer's thinking mode is the two-valued logic—one or the other. But the human brain is quite different. In the cerebral cortex, there are more than 3000 synapses in a pyramidal cell. A synapse has numerous ion channels. When the information passes from one cell to another, it has passed many ion channels in which there are numerous intermediate states from totally opening to completely closing. Thus, the synapse opens different ion channels to different synaptic pink shades, because the different ion channels are stimulated to open. The different ion channels of the synapse "switch on or off" accordingly just as the synapse partially opens or closes the door in different positions. That is to say, if the number is used to express these, in addition to these two extremes: the fully closing—0, and the fully opening—1, it can correspond to any decimals from 0 to 1, so the human brain can distinguish different shades of pink. Further speaking, the human brain can detect all different degrees of fuzziness. Thus, the thinking mode of the human brain is not a simple binary logic (black and white, yes and no, right and wrong …), but more full-scale and more realistic fuzzy logic (black–dark gray–gray–gray white–white; specious; entirely correct–basically correct–half correct and half wrong–a little wrong–wrong–basically wrong–completely wrong …). Apparently, the former depends on traditional precise mathematics, and the latter needs to create a new science to research fuzziness—Fuzzy mathematics, in order to imitate the human brain's fuzzy logic thinking.

1.2.6   Birth of Fuzzy Mathematics

1.2.6.1   How to Teach a Computer to Recognize Handwriting: *a* and *d*—Found by Prof. Zadeh

How could we make the computer think and judge like the human brain? Scientists were thinking about it, but mathematicians did not know about these findings of human brain that biologists have found, and they were still researching agonizingly: When a computer judges a person who is walking here from a distance, you need to enter the person's tallness, weight, stride, arm swing frequency, and the angular speed into the computer... Even though human beings still fail to know exactly a person, they are able to identify the person far away as long as they are familiar with 'his' body and walking posture, which can be compared with the stored information in their brain.... People can do this in fact based on a general impression, which do not need very precise data, but only the enough fuzzy information. For another example, when we use fuzzy concept to describe an object in a particular environment, we can understand each other. What does it mean to say that? Generally it will not cause misunderstanding and ambiguity. In case, someone in a hurry wants to find a colleague at a meeting while you are going just into the conference room. Then he asks you to call the person out. He tells you that this person is characterized by the bald, heavy beard, thick eyebrows, and big eyes. As long as you enter the room and sweep away, you will find this person. But if he told you exactly that the person had 12,368 hairs and 4821 beards, and exact eye size, you might not be able to judge who he is. In this case, given the fuzzy concept, you will get the accurate result; on the contrary, if given a precise description, the result will be blurred. This is because the vague concept can cover a broader range than a precise description. This qualitative description can use the people's existing knowledge to increase greatly the amount of information. The human language is fuzzy, which reflects the way people think. We all can understand and never puzzle the difference between the people's thinking mode and the traditional mathematics. We can think further that accurateness is hidden by fuzziness; fuzziness is supported by accurateness. Some scholars have been thinking and studying how to describe the blurring of the objective world since 1920. In fact, it is the difference of all things in the world that produces fuzzy mathematics because the difference is not one or the other; yes, or no; but there are infinite intermediate transitional states, gradually evolved. So the original two values of 0,1 in the computer are not enough to represent the things. They must be extended to any values from 0 to 1. With the values lessened from 1 onwards (quantitatively changing), it would come to be changed from one thing to another (qualitatively changed), such as young → not young → old, white → gray → gray black → black etc. Fuzzy mathematics was born in order to analyze and deal with the ambiguity of things everywhere. In 1965, Professor L. A. Zadeh at University of California, who had long been engaged in Automatic Control Research, found that the human brain can discriminate English letters according to their similar level between some handwriting and standard mode (as the difference of handwritten letters between a and d depends on "how long the vertical bar is on a"), while he was making researches on the machine recognition of handwritten letters of the alphabet. Obviously the thinking process of the human brain would have fuzziness, by which he completed his first paper on fuzzy mathematics defined as "Fuzzy sets".

Let us first review what the concept of a common set is.

1.3   Sets

1.3.1   Concept and Characteristics

The sets refer to a collection with a certain collective property of the object.

Features: (1) disorder; (2) mutual exclusion; (3) certainty.

1.3.2   Representation of Sets

(1) **Listing Method**

Example 1.1 Set A is supposed for the municipalities of our country. Then A = {Beijing, Shanghai, Tianjin, Chongqing}.

(2) **Description Method (or Definition Method)**

Example 1.2 Set B is for positive even numbers, then $B = \{x|x = 2n, n \in N\}$.

To introduce fuzzy sets, the third set presentation is explained as follows:

(3) **Representation of Characteristic Function**

For a common set A, the element X either belongs to A, or does not belong to A, this or that, is very clear. We define a function as:

$$f_A(x) = \begin{cases} 1, & x \in A, \\ 0, & x \notin A, \end{cases}$$

$f_A(x)$ is called characteristic function of set A, which characterizes the elements' membership conditions of set A. The value 1 or 0 of the characteristic function is called membership degree. When $x_1 \in A$, x_1's membership degree $f_A(x_1) = 1$, i.e., x_1 belongs to set A; when $x_2 \notin A$, x_2's membership degree $f_A(x_2) = 0$ i.e., x_2 does not belong to set A.

Example 1.3 A = {all the boys of Class (1) Senior 2}

$$f_A \text{ (Gang Wang)} = 1,$$
$$f_A \text{ (Zhen Wang)} = 0,$$
$$f_A \text{ (Tie Li)} = 0.$$

Obviously, Gang Wang is a boy of Class (1) Senior 2, belongs to set A. Zhen Wang is a girl, and Tie Li is not a boy of Class (1) Senior 2, both do not belong to set A.

Example 1.4 X = {x_1, x_2, x_3, x_4, x_5} denotes a set of five students. Evaluation set A: Good in math. In terms of the general view of common sets, select a characteristic function

$$f_A(x) = \begin{cases} 1, & x \in A \text{---(good at math)}, \\ 0, & x \notin A \text{---(bad at math)}. \end{cases}$$

If $f_A(x_1) = f_A(x_3) = f_A(x_5) = 1, f_A(x_2) = f_A(x_4) = 0$, then A = {1,0,1,0,1}. It is impossible to separate the difference of the five students. It is because the "good" or the "bad" is a vague concept, only two values 1 and 0 can not tell the subtle difference between them. It need improve further, so, we introduce fuzzy sets.

1.4 Fuzzy Sets

1.4.1 Concepts

All elements have some nature; these elements do not have some clear boundaries. Such a set is called fuzzy set, denoted by the symbols $\underset{\sim}{A}$, $\underset{\sim}{B}$ etc.

1.4.1.1 Visual Images of Sets and Fuzzy Sets

We can use a visual image to explain the concept of a set. In Domain U, element μ can be imagined as a "dot" without any size and quality in the U. Set A in the U can be regarded as in a "circle" (Fig. 1.8). When the dot μ is inside the circle, and its

Fig. 1.8 A domain, a set, and elements

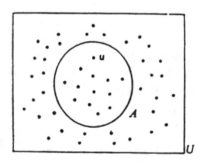

Fig. 1.9 The visual image of
a fuzzy set

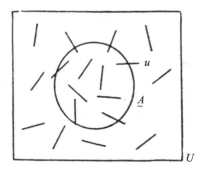

characteristic function's value $f_A(\mu) = 1$. The element μ belongs to set A; When dot μ is outside the circle, and its characteristic function's value $f_A(\mu) = 0$. The element μ does not belong to set A.

Similarly, you can use a visual image to explain the concept of a fuzzy set.

We now change the above "dot" in Domain U into a line segment with unit length of 1, see Fig. 1.9. In this case, we record a fuzzy set by $\underset{\sim}{A}$. If element μ is inside an $\underset{\sim}{A}$ circle, denoted by 1; If the μ is outside the $\underset{\sim}{A}$ circle, denoted by 0; If the μ is partly inside the $\underset{\sim}{A}$ and partly outside the $\underset{\sim}{A}$, the length of the μ inside the $\underset{\sim}{A}$ is presented by the membership degree of μ to $\underset{\sim}{A}$. The longer the length is, the more membership degree it will be.

Can we also create a function to characterize a fuzzy set? The problem is not "whether the line segment is inside the μ circle or outside it". It is "how long the line segment inside the μ is".

Obviously, the function value represented by only 0, 1 is not enough. According to our above analysis of the "synapses", the range of the function value should be expanded to the closed interval [0, 1].

Example 1.5 The "tall" is a vague concept, how much is tallness? 1.70 m? 1.75 m? 1.80 m? or 1.85 m?

This is a proposition related to races, nutrition, and ages. If the function value of 1 expresses certain tallness and 0 certain shortness, 0.8 denotes the taller and 0.3 the shorter etc.

Similarly, in Example 1.4, if we use percentile achievements as the basis of "good" and "bad", Suppose 100 is "good", the function value takes 1; 75 is "better", the function value 0.75; 50 is "not good", the function value 0.5; and so on ..., by which we have the average marks of the five students in the last semester as 95, 56, 90, 63, 85, $\underset{\sim}{A} = \{0.95, 0.56, 0.90, 0.63, 0.85\}$, available to reflect the difference of the five students.

1.4.1.2 Searching with Keywords is Fuzzy Technology

The most widely used case in fuzzy concept is to retrieve. The so-called keyword technology is the fuzzy technology. Before the 1980s, if you called 114 for a company's telephone number, the operator would ask you to tell his or her company's full name. One word more or less, it would not be found because it is based on the traditional math, absolutely precise at that time. In most cases, people tend to remember the general name. For example, when we go to a library to borrow some books, one word more or less of the book's name is often too vague to remember. With the keyword technology, it is helpful. We can find out all the books containing the word and list them according to the closeness of the word order. The more you remember to the book's keywords, the topper its name will be listed.

Below, let us see how to use a membership function to make the fuzzy concept digital.

1.4.2 Membership Function and Membership Degree

1.4.2.1 Sets—Characteristic Function; Fuzzy Sets—Membership Function

For fuzzy set $\underset{\sim}{A}$, define a function $\mu_A(x)$, for any element x, it has $0 \le \mu_A(x) \le 1$, namely $\mu_A(x) \in [0, 1]$. We call it a membership function. The membership function $\mu_A(x)$ determines a fuzzy set $\underset{\sim}{A}$. The value of $\mu_A(x)$ is called membership degree of x to $\underset{\sim}{A}$. The closer the value is to 1, the higher the membership degree of x to $\underset{\sim}{A}$ is. The closer the value is to 0, the lower the membership degree of x to $\underset{\sim}{A}$.

Take the fuzzy set of "young" as an example, there are four young people in an office. Xiao Zhang is 19 years old, the membership degree of him to "young" μ_A (Xiao Zhang) = 0.9. Xiao Li is 28 years old, μ_A (Xiao Li) = 0.46. Xiao Wang has worked for years, and is 34 years old; the membership degree of him to "young" is closer to 0. Xiao Zhao just came here, the youngest, is only 18 years old, the membership degree of him to "young" is closer to 1. The value of the membership function accurately reflects the age of the young: Xiao Zhao is the youngest; Xiao Zhang was very young; Xiao Li is younger; and Xiao Wang is not too young. Here, the words: "youngest", "very", "younger", and "not too" etc. indicate the degree of natural language, which are converted into precise numbers through the membership function. In this way, the computer can understand and accept the fuzzy phenomena.

In order to understand the difference between a common set and a fuzzy set by the visual picture, we compare the curve of a characteristic function with that of membership function. In Fig. 1.10, the right characteristic function shows a temperature set of water frozen to ice; The horizontal axis represents the temperature of the water, ranging from −10 to 10 °C; and the vertical axis is the value of

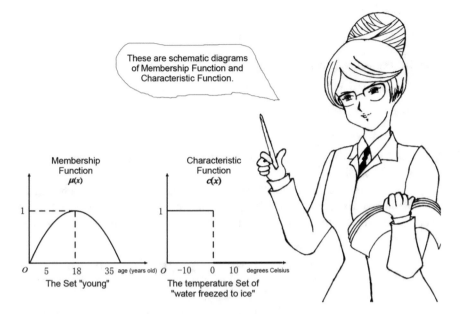

Fig. 1.10 Membership function and characteristic function

characteristic function. From natural phenomena, we know that water freezes at zero degree Celsius or less, and does not freeze at above it. The zero degree Celsius in the characteristic function is a dividing line. Not more than zero degree Celsius, the value of characteristic function is equal to 1; higher than zero degree Celsius, the value of characteristic function is equal to 0. To see from its curve, the value of the function from 1 jumping down to 0 at the demarcation point of zero degree Celsius. The membership function does not exist as such a cut-off point, and the function curve also changes continuously without any sudden jump in. It actually reflects the characteristics of fuzzy set without any distinct border.

1.4.3   Representation of Fuzzy Sets

1. **Finite Domain**

When X is a finite domain, a fuzzy set $\underset{\sim}{A}$ in X usually has three kinds of representations.

(a) **Zadeh Representation**

Professor Zadeh expresses the membership degree of every element of a fuzzy set by fraction, i.e., a fuzzy set $\underset{\sim}{A}$ is a finite set $\{x_1, x_2, \ldots, x_n\}$ then it can be represented by a general formula as follows:

$$A = \frac{\mu_1}{x_1} + \frac{\mu_2}{x_2} + \cdots + \frac{\mu_n}{x_n},$$

In the formula, μ_i is the abbreviation for $\mu_A(x_i)$, showing the membership degree of No. i element to A. $\frac{\mu_1}{x_1}$ does not represent a fraction, but the corresponding relation of the membership degree μ_i of x_i to A; "+" does not represent the addition, but the whole fuzzy set in domain X.

Example 1.6 $X = \{1, 2, 3, 4, 5, 6, 7, 8, 9, 10\}$ as a domain, let us discuss the fuzzy concept of "several".

The fuzzy concept is the fuzzy set in mathematical language. Although its connotation and denotation is not clear, it can quantitatively make the membership function, based on experiences, the fuzzy set of "several" can be expressed as follows:

$$[\text{Several}] = \frac{0}{1} + \frac{0}{2} + \frac{0.3}{3} + \frac{0.7}{4} + \frac{1}{5} + \frac{1}{6} + \frac{0.7}{7} + \frac{0.3}{8} + \frac{0}{9} + \frac{0}{10}.$$

In the above equation, the membership degree of five or six is 1, indicating that it is of the most possibility that we would use "several" to express five or six; it is of seventy percent of possibility that we would use "several" to express four or seven; it is of thirty percent of possibility that we would use "several" to express three or eight; and in Chinese we seldom use "several" to express one, two or nine, ten (expressing it, we use other words, such as "one or two", or "eight or nine", "by ten", "dozen" etc.), so their membership degree is zero.

If we give up those with a zero membership degree, then:

$$[\text{Several}] = \frac{0.3}{3} + \frac{0.7}{4} + \frac{1}{5} + \frac{1}{6} + \frac{0.7}{7} + \frac{0.3}{8}.$$

(b) **Ordered Pair Representation**

$$A = \{(x_1, \mu_1), (x_2, \mu_2), \ldots, (x_n, \mu_n)\}.$$

As in Example 1.6, it can also be expressed as follows:

$$[\text{Several}] = \{(1, 0), (2, 0), (3, 0.3), (4, 0.7), (5, 1), (6, 1), (7, 0.7), (8, 0.3), (9, 0), (10, 0)\}.$$

(c) **Vector Representation**

$$A = \{\mu_1, \mu_2, \ldots, \mu_n\}.$$

As in Example 1.6, it still can be expressed below:

$$[\text{Several}] = \{0, 0, 0.3, 0.7, 1, 1, 0.7, 0.3, 0, 0\}.$$

2. **Infinite Domain**

Function Representation

When X is the representation of an infinite domain:
    If there are infinite elements in a Set A, A can be expressed by borrowing an integral symbol as follows:

$$A = \int_{A} \mu_A(x)/x.$$

The integral symbol $\int$ here does not represent the integral calculus, whereby it only represents an infinite number of combined elements, representing a summary of the corresponding relation of the membership degree μ of x in domain X.
    As a fuzzy set of "old", it also can be given by:

$$[\text{Old}] = \int_{50}^{200} \left[1 + \left(\frac{x-50}{5}\right)^{-2}\right]^{-1} \Bigg/ x.$$

The subscript on the integral symbol represents that 50 (years old) is the lowest limit of the "old" fuzzy set. People less than 50 years old do not belong to the "old" set; the superscript on the integral symbol represents that 200 (years old) is the upper limit of the set. The fraction after the integral symbol still represents the membership degree of a certain element x. In this case, it means that someone belongs to the old people, such as:

$$\mu_{\text{Old}}(51) = \frac{1}{26} = 0.04 \text{---} \mu_{\text{Old}}(55) = \frac{1}{2} = 0.5,$$

$$\mu_{\text{Old}}(65) = \frac{9}{10} = 0.9 \text{---} \mu_{\text{Old}}(71) = \frac{1}{1.057} = 0.95. \ldots$$

Example 1.7 In set H of "tall", if we set 1.80 m as a tall person, and 1.50 m as a dwarf, we can roughly define the membership function $\mu_H(x)$ by the visual image as follows (Fig. 1.11):

Fig. 1.11 The function image of "tall"

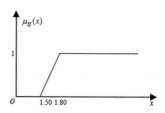

$$\mu_H(x) = \begin{cases} 1, & x \geq 1.80 \text{ m}, \\ \dfrac{1}{1.80 - 1.50}(x - 1.50), & 1.50 \text{ m} < x < 1.80 \text{ m}, \\ 0, & x \leq 1.50 \text{ m}. \end{cases}$$

With the linear increasing features, we can use the slope to list the function expression of the middle section $\dfrac{1}{1.80 - 1.50}(x - 1.50)$.

Example 1.8 Try to complete the membership function of set Y of "young", set from 15 to 25 years old for young people, and the age of either more than 35 or less than 10 years old is not for young people. It means a trapezoidal image. Now we can come to the following image according to the above principle (Fig. 1.12).

Fig. 1.12 The function image of "young"

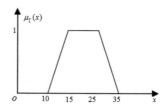

$$\mu_y(x) = \begin{cases} 0, & x < 10, \\ \dfrac{1}{15 - 10}(x - 10), & 10 \leq x < 15, \\ 1, & 15 \leq x \leq 25, \\ \dfrac{1}{25 - 35}(x - 35), & 25 < x < 35, \\ 0, & x \geq 35. \end{cases}$$

Example 1.9 Seek the membership function $\mu_S(x)$ of Set $\underset{\sim}{S}$ of "the number near 4" (Fig. 1.13):

Fig. 1.13 The function image of $\mu_s(x)$

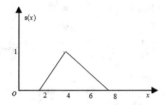

$$\mu_s(X) = \begin{cases} 0, & x \le 2, \\ \dfrac{1}{4-2}(x-2), & 2 < x < 4, \\ 1, & x = 4, \\ \dfrac{1}{4-6}(x-6), & 4 < x < 6, \\ 0, & x \ge 6. \end{cases}$$

1.4.4 Determination of Membership Functions

In fact, the determination and solution to the membership function is not unique because the blurred boundaries in everyone's mind are not totally the same. Although in essence, a membership function is an objective quantitative characterization, the establishment of membership function allows certain man's skills. It allows flexible construction based on everyone's professional knowledge and practical experiences without uniform models. In spite of the different forms of membership functions, they can reach the same goal by different routes as long as they can aptly portray with the fuzzy phenomena solved and dealt. In practice, the following methods are commonly used such as a fuzzy statistical test, function segmentation, binary comparison taxis, and finding membership functions by operation among the fuzzy sets and so on. Here we only make a brief introduction to the fuzzy statistical method.

1.4.4.1 Fuzzy Statistics

This method can generally make a fuzzy statistical test for an amount of people. Making frequency histograms first, and delineating the images continuously, we can obtain a membership function curve. Then by the curve fitting method, we can obtain a membership function. For example, the age domain $U = [0,100]$ (unit: years old). $\underset{\sim}{A}$ is a fuzzy set in the U, representing fuzzy concept α = youth. Select the age $\mu_0 = 27$, try to use the fuzzy statistical test to determine the membership degree of μ_0 to $\underset{\sim}{A}$.

Table 1.1 The age range of "youth"

Group	1	2	3	4	5	6	7	8	9	10
No. 1	18–25	17–30	17–28	18–25	16–35	14–25	18–30	18–35	18–35	16–25
No. 2	15–30	18–35	17–30	18–25	10–25	18–35	20–30	18–30	16–30	20–35
No. 3	18–30	18–30	15–25	18–30	15–28	16–28	16–30	18–30	16–30	18–35
No. 4	18–25	18–25	16–28	18–30	16–30	16–28	18–35	18–35	17–27	16–28
No. 5	15–28	16–30	19–28	15–30	15–26	17–25	15–36	18–30	17–30	18–35
No. 6	16–35	15–25	15–25	18–28	16–30	15–28	18–35	18–30	17–28	18–35
No. 7	15–28	18–30	15–25	15–25	18–30	16–24	15–25	16–32	15–27	18–35
No. 8	16–25	18–28	16–28	18–30	18–35	18–30	18–30	17–30	18–30	18–35
No. 9	16–30	18–35	17–25	15–30	18–25	17–30	14–25	18–26	18–29	18–35
No. 10	18–28	18–30	18–25	16–35	17–29	18–25	17–30	16–28	18–30	16–28
No. 11	15–30	15–35	18–30	20–30	20–30	16–25	17–30	15–30	18–30	16–30
No. 12	18–28	18–35	16–30	15–30	18–35	18–35	18–30	17–30	18–35	17–30
No. 13	15–25	18–35	15–30	15–25	15–30	18–30	17–25	18–29	18–28	

Table 1.2 The membership frequency $\mu_0 = 27$ to the age of "youth"

The first n persons	10	20	30	40	50	60	70	80	90	100	110	120	129
Membership degree	6	14	23	31	39	47	53	62	68	76	85	95	101
Membership frequency	0.60	0.70	0.77	0.78	0.78	0.78	0.76	0.78	0.76	0.76	0.75	0.79	0.78

Professor Nanlun ZHANG in Wuhan Institute of Building Materials made a sampling test. He selected 129 suitable candidates. They reported what the most appropriate age for "youth" should be after their own serious consideration of the meaning of "youth". See Table 1.1 for the specific data. See Table 1.2 for the membership frequency $\mu_0 = 27$ to the age of "youth".

Obviously, the membership frequency $\mu_0 = 27$ to the age of "youth" is broadly at the stable frequency of near 0.78. So we have $\mu_A(27) = 0.78$.

In addition, it is easy to find a membership function of the youth A. Separate U into groups, calculate the membership frequency by the median as representing values of each group, see Table 1.3.

According to Table 1.3, we can delineate histograms (Fig. 1.14), and then delineate the images continuously to obtain the membership function curve $\mu_A(u)$ (see Fig. 1.15). Of course, we can use the curve fitting method for the membership function.

Table 1.3 The membership frequency

No.	Group	Frequency	Relative frequency
1	13.5–14.5	3	0.0232
2	14.5–15.5	27	0.2093
3	15.5–16.5	51	0.3953
4	16.5–17.5	67	0.5194
5	17.5–18.5	124	0.9612
6	18.5–19.5	125	0.9690
7	19.5–20.5	129	1.000
8	20.5–21.5	129	1.000
9	21.5–22.5	129	1.000
10	22.5–23.5	129	1.000
11	23.5–24.5	129	1.000
12	24.5–25.5	128	0.9922
13	25.5–26.5	103	0.7984
14	26.5–27.5	101	0.7829
15	27.5–28.5	99	0.7674
16	28.5–29.5	80	0.6202
17	29.5–30.5	77	0.5969
18	30.5–31.5	27	0.2093
19	31.5–32.5	27	0.2093
20	32.5–33.5	26	0.2016
21	33.5–34.5	26	0.2016
22	34.5–35.5	26	0.2016
23	35.5–36.5	1	0.0078
Σ			13.6589

Professor Nan-lun ZHANG and other professors have made statistical tests for the membership function curve of the "youth" (to 106 people) in Department of Biology, Wuhan University, (to 129) in Wuhan Institute of Building Materials, and (to 94) in Xi'an Industrial College. The results of three statistical curves have almost the same shape and the same total area enclosed under the curve. This shows that the fuzzy statistical tests also have a characteristic of membership frequency stability (see Chap. 4 Fuzzy Statistics for details).

With commonly nine kinds of curves, expressions for membership functions as follows (see the images at the next page), where a, k are as parameters, derived the values from a large number of statistical data. Since the determination of membership functions involves a lot of knowledge of higher mathematics, we can only give a rough introduction here.

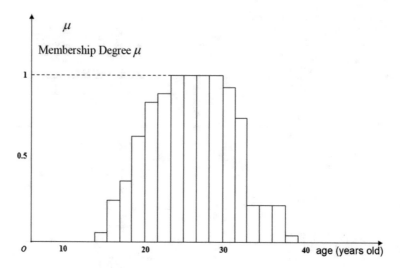

Fig. 1.14 The membership function histograms of the "youth"

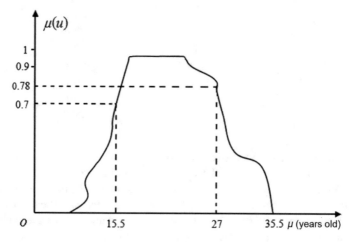

Fig. 1.15 The membership function curve of $\mu_{\underset{\sim}{A}}(u)$

① Partial Small Type (Supremum)

i. Fall Semi- γ -
Distribution

$$\mu_{\underset{\sim}{A}}(x) = \begin{cases} 1, & x \le a, \\ e^{-k(x-a)}, & x > a. \end{cases}$$

where $k > 0$.

ii. Fall Semi-Normal
Distribution

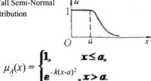

$$\mu_{\underset{\sim}{A}}(x) = \begin{cases} 1, & x \le a, \\ e^{-k(x-a)^2}, & x > a. \end{cases}$$

where $k > 0$.

iii. Fall Semi-
Trapezoidal
Distribution

$$\mu_{\underset{\sim}{A}}(x) = \begin{cases} 1, & x \le a_1, \\ \dfrac{a_2 - x}{a_2 - a_1}, & a_1 < x \le a_2, \\ 0, & x > a_2. \end{cases}$$

② Partial Large Type (Infimum)

i. Rise Semi- γ -
Distribution

$$\mu_{\underset{\sim}{A}}(x) = \begin{cases} 0, & x \le a, \\ e^{-k(x-a)}, & x > a. \end{cases}$$

where $k > 0$.

ii. Rise Semi-
Normal Distribution

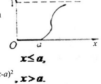

$$\mu_{\underset{\sim}{A}}(x) = \begin{cases} 0, & x \le a, \\ e^{-k(x-a)^2}, & x > a. \end{cases}$$

where $k > 0$.

iii. Rise Semi-
Trapezoidal
Distribution

$$\mu_{\underset{\sim}{A}}(x) = \begin{cases} 0, & x \le a_1, \\ \dfrac{a_2 - x}{a_2 - a_1}, & a_1 < x \le a_2, \\ 1, & x > a_2. \end{cases}$$

③ Medial Type (Symmetric Type)

i. Sharp γ -
Distribution

$$\mu_{\underset{\sim}{A}}(x) = \begin{cases} e^{k(x-a)}, & x \le a, \\ e^{-k(x-a)}, & x > a. \end{cases}$$

where $k > 0$.

ii. Normal
Distribution

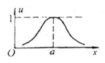

$$\mu_{\underset{\sim}{A}}(x) = e^{-k(x-a)^2}.$$

where $k > 0$.

III. TRAPEZOIDAL
DISTRIBUTION

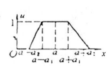

$$\mu_{\underset{\sim}{A}}(x) = \begin{cases} 0, & x \le a - a_2, \\ \dfrac{a_2 + x - a}{a_2 - a_1}, & a - a_2 < x \le a - a_1, \\ 1, & a - a_1 < x \le a + a_1, \\ \dfrac{a_2 - x + a}{a_2 - a_1}, & a + a_1 < x \le a + a_2, \\ 0, & x > a + a_2. \end{cases}$$

1.4.5   Operations of Fuzzy Sets

We have learned basic operations of common sets, such as relation of combining, intersection, complement, and containing. We still need to make other definitions for basic operations of fuzzy sets because of an extension of fuzzy sets in common sets.

(1) **Equality**:

Since the feature of a fuzzy set is its membership function, naturally two fuzzy sets can be defined as equal if all their membership functions appear the same. For all elements x in the domain, if

$$\mu_A(x) = \mu_B(x),$$

then $A = B$.

Similarly, complement set $\overline{A}$ of fuzzy set A can be defined by operations of the membership function $\mu_A(x)$.

(2) **Complement:**

If the following equation is correct,

$$\mu_A(x) = 1 - \mu_A(x),$$

we can call fuzzy set $\overline{A}$ of the above a membership function, i.e., complement set of A. It also can be referred to as A^c.

For example, A is on behalf of the "tall" fuzzy set, $\overline{A}$ is the "non-tall" fuzzy set, and x_1 is 1.78 m tall. If $\mu_A(x_1) = 0.9$, then the "non-tall" qualification is:

$$\mu_A(x_1) = 1 - 0.9 = 0.1.$$

In addition, when A is a common set

$$\mu_A(x) = 1, \quad then \quad \mu_{\overline{A}}(x) = 0,$$
$$\mu_A(x) = 0, \quad then \quad \mu_{\overline{A}}(x) = 1.$$

This is entirely consistent with the definition as the complement sets of fuzzy sets.

Similarly, for all x, if $\mu_{\underset{\sim}{A}}(x) = 0$, then $\underset{\sim}{A}$ can be defined as an empty fuzzy set, denoted by $\varnothing$. Contrarily, if for all x, there is $\mu_{\underset{\sim}{A}}(x) = 1$, called a full set $\underset{\sim}{A}$. The empty set and the full set are of mutual complement sets.

(3) **Containing**:

For all elements x, if $\mu_{\underset{\sim}{B}}(x) \leq \mu_{\underset{\sim}{A}}(x)$, calling that fuzzy set $\underset{\sim}{A}$ contains fuzzy set $\underset{\sim}{B}$, denoted as

$$\underset{\sim}{B} \subset \underset{\sim}{A}.$$

Then, $\underset{\sim}{B}$ is the fuzzy subset of $\underset{\sim}{A}$.

Example 1.10 Domain $X = \{a, b, c\}$, $\underset{\sim}{A}$, $\underset{\sim}{B}$ are two fuzzy subsets. There exist

$$\mu_{\underset{\sim}{B}}(a) = 0.6, \quad \mu_{\underset{\sim}{A}}(a) = 1,$$
$$\mu_{\underset{\sim}{B}}(b) = 0, \quad \mu_{\underset{\sim}{A}}(b) = 0.4;$$
$$\mu_{\underset{\sim}{B}}(c) = \mu_{\underset{\sim}{A}}(c) = 1;$$
$$\therefore \underset{\sim}{B} \subset \underset{\sim}{A}.$$

Another example, $\underset{\sim}{A}$ denotes the fuzzy set of "smart people", $\underset{\sim}{B}$ represents the fuzzy set of "very smart people." The membership degree of anyone who belongs to $\underset{\sim}{B}$ is always less than that of anyone who belongs to $\underset{\sim}{A}$. Therefore, "very smart people" is the fuzzy subset of "smart people" while "smart people" contains "very smart people".

(4) **Union**—Take the Bigger

For union $\underset{\sim}{C}$ of two fuzzy sets $\underset{\sim}{A}$ or $\underset{\sim}{B}$, their membership functions are defined as

$$\mu_{\underset{\sim}{C}}(x) = \max\left[\mu_{\underset{\sim}{A}}(x), \mu_{\underset{\sim}{B}}(x)\right],$$

it can also be expressed as

$$\mu_{\underset{\sim}{C}}(x) = \mu_{\underset{\sim}{A}}(x) \vee \mu_{\underset{\sim}{B}}(x),$$

where the symbol "∨" means to take the bigger operation, i.e., taking the bigger number of both sides of "∨" to calculate the result. It is expressed by the set symbols:

$$C = A \cup B.$$

If A or B are for common sets, as defined by the common set theory, combining set C is at least the set made up of the elements belonging to A or B. In the above definition of fuzzy sets and union operations, if the membership function only takes 1 or 0, then it becomes the union operations of common sets.

E.g., A = {m, n, v}, B = {m, p}, it can obtain

$$\mu_A(m) = 1 \text{---} \mu_B(m) = 1,$$
$$\mu_A(n) = 1 \text{---} \mu_B(n) = 0,$$
$$\mu_A(v) = 1 \text{---} \mu_B(v) = 0,$$
$$\mu_A(p) = 0 \text{---} \mu_B(p) = 1.$$

From "taking the bigger" operation, we get

$$\mu_c(m) = \mu_A(m) \vee \mu_B(m) = 1 \vee 1 = 1;$$
$$\mu_c(n) = \mu_A(n) \vee \mu_B(n) = 1 \vee 0 = 1.$$

Similarly,

$$\mu_c(v) = 1, \quad \mu_c(p) = 1.$$
$$\therefore A \cup B = C = \{m, n, v, p\}.$$

This operation result is exactly the same as that of common sets. Therefore, a common set is just a special case of a fuzzy set. The Union operation of Fuzzy Sets is the expansion and generalization of that of Common Sets.

Example 1.11 A set is composed by the five persons $X = \{x_1, x_2, x_3, x_4, x_5\}$. Try to investigate: Set A of "tall persons" and set B of "fat persons" if the membership functions of A and B are respectively as follows:

$$\begin{cases} \mu_A(x_1) = 0.6, \\ \mu_A(x_2) = 0.5, \\ \mu_A(x_3) = 1, \\ \mu_A(x_4) = 0.4, \\ \mu_A(x_5) = 0.3, \end{cases} \qquad \begin{cases} \mu_B(x_1) = 0.5, \\ \mu_B(x_2) = 0.6, \\ \mu_B(x_3) = 0.3, \\ \mu_B(x_4) = 0.4, \\ \mu_B(x_5) = 0.7. \end{cases}$$

At this time, $(A \text{ or } B)$'s union $C = A \cup B$, representing fuzzy set of "tall or fat persons", the membership functions are

$$\begin{cases} \mu_{\underset{\sim}{c}}(x_1) = 0.6, \\ \mu_{\underset{\sim}{c}}(x_2) = 0.6, \\ \mu_{\underset{\sim}{c}}(x_3) = 1, \\ \mu_{\underset{\sim}{c}}(x_4) = 0.4, \\ \mu_{\underset{\sim}{c}}(x_5) = 0.7. \end{cases}$$

We can see x_3, x_5 are better to be in line with requirements of C.

Also, because

$$\mu_A(x) \leq \max[\mu_A(x), \mu_B(x)],$$
$$\mu_B(x) \leq \max[\mu_A(x), \mu_B(x)],$$

according to the containing definition in fuzzy sets, we can see

$$A \subset A \cup B,$$
$$B \subset A \cup B,$$

A and B are contained in their union.

(5) **Intersection**—take the smaller

Similarly, to define the intersection of fuzzy sets A and B, contrary to the case of union, we should select the smaller of the two membership functions as the membership function of intersection. $(A \text{ and } B)$'s intersection is referred to as

$$A \cap B,$$

its membership function is defined as

$$\mu_{A \cap B}(x) = \min\left[\mu_A(x), \mu_B(x)\right].$$

The formula above can also be expressed as:

$$\mu_{A \cap B}(x) = \mu_A(x) \wedge \mu_B(x).$$

where the symbol "∧" indicates to take a smaller operation, i.e., taking the smaller number at both ends of "∧" as a result of the operation.

Such as for Example 1.11 A set of five persons may have

$$
\begin{cases}
\mu_{A \cap B}(x_1) = 0.5, \\
\mu_{A \cap B}(x_2) = 0.5, \\
\mu_{A \cap B}(x_3) = 0.3, \\
\mu_{A \cap B}(x_4) = 0.4, \\
\mu_{A \cap B}(x_5) = 0.3.
\end{cases}
$$

Here, the intersection $A \cap B$, represents fuzzy set of "tall and fat persons", only x_1, x_2 barely meet the requirements.

For the intersection, because

$$\mu_A(x) \geq \min\left[\mu_A(x), \mu_B(x)\right],$$

$$\mu_B(x) \geq \min\left[\mu_A(x), \mu_B(x)\right],$$

then

$$A \supset A \cap B,$$
$$B \supset A \cap B.$$

In full set X, the membership functions of every element $\mu_A(x), \mu_B(x)$, $\mu_{A \cup B}(x), \mu_{A \cap B}(x)$ are shown in Figs. 1.16 and 1.17 respectively.

Fig. 1.16 $A \cup B$

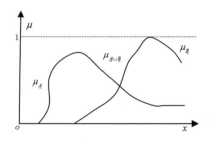

Fig. 1.17 $A \cap B$

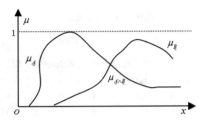

Similarly, the intersection definition for fuzzy sets is also the extension of that for common sets. As long as the membership functions take only 1 or 0, the intersection of fuzzy sets will become that of common sets.

The union and intersection operations of fuzzy sets are not only as similar as those of the same kinds of common sets, but also have practical significance. For example, in a unified entrance examination for graduate students, two foreign languages of English and Japanese were taken. If a candidate gets 90 marks in English and 60 in Japanese, it can take 90 marks on behalf of the student's foreign language proficiency (Take the bigger) this time if some supervisors only require candidates to master a foreign language; or it only can take 60 marks on behalf of the student's foreign language proficiency (Take the smaller) if some supervisors require candidates to master two foreign languages.

Many natures in the common Sets' operations, such as the law of commutation and so on, tend to be suitable for that of the combination and intersection of fuzzy sets, but not all natures are suitable. The Law of Excluded Middle is undermined. For example

$$A \cap \overline{A} = \varnothing,$$
$$A \cup \overline{A} = U.$$

The above are not necessarily suitable for fuzzy sets.

As in Example 1.11, the "tall" set $A = (0.6, 0.5, 1, 0.4, 0.3)$, but

$$\overline{A} = (1 - 0.6, 1 - 0.5, 1 - 1, 1 - 0.4, 1 - 0.3)$$
$$= (0.4, 0.5, 0, 0.6, 0.7),$$

thereby

$$A \cap \overline{A} = (0.6 \wedge 0.4, 0.5 \wedge 0.5, 1 \wedge 0, 0.4 \wedge 0.6, 0.3 \wedge 0.7)$$
$$= (0.4, 0.5, 0, 0.4, 0.3) \neq \varnothing,$$
$$A \cup \overline{A} = (0.6 \vee 0.4, 0.5 \vee 0.5, 1 \vee 0, 0.4 \vee 0.6, 0.3 \vee 0.7)$$
$$= (0.6, 0.5, 1, 0.6, 0.7) \neq U.$$

A's complement set $\overline{A}$, logically, it is the negative proposition of the proposition. $A \cap \overline{A} \neq \varnothing$ can be understood as a proposition and its negative proposition that are not entirely mutually exclusive. The result is absurd in terms of traditional mathematics. However, think carefully about it in life. We can indeed encounter this situation: someone with about 1.7 m tall, we say he is not tall yet. We do not say he is tall, but do not mean that he is short. So in terms of fuzzy sets, it is not "either this or that" but allows "both this and that". Getting rid of Law of Excluded Middle is to get rid of the absolutization of the membership relation. It is the fuzzy set theory that makes the logic progress in line with people's thinking law.

Example 1.12 As is shown in Fig. 1.18a–e is 5 pieces; they are composed of the domain:

$$A = \text{"circle"} = \frac{1}{a} + \frac{0.75}{b} + \frac{0.5}{c} + \frac{0.25}{d},$$

$$B = \text{"square"} = \frac{0.3}{b} + \frac{0.5}{c} + \frac{0.7}{d} + \frac{1}{e},$$

Fig. 1.18 $X = \{a, b, c, d, e\}$

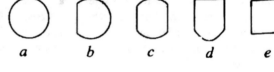

$$a \qquad b \qquad c \qquad d \qquad e$$

$A \cup B = \text{"square or circle"}$

$$= \frac{1 \vee 0}{c} + \frac{0.75 \vee 0.3}{b} + \frac{0.5 \vee 0.5}{c} + \frac{0.25 \vee 0.7}{d} + \frac{0 \vee 1}{e}$$

$$= \frac{1}{a} + \frac{0.75}{b} + \frac{0.5}{c} + \frac{0.7}{d} + \frac{1}{e},$$

$A \cap B = \text{"square and circle"}$

$$= \frac{1 \wedge 0}{c} + \frac{0.75 \wedge 0.3}{b} + \frac{0.5 \wedge 0.5}{c} + \frac{0.25 \wedge 0.7}{d} + \frac{0 \wedge 1}{e}$$

$$= \frac{0.3}{b} + \frac{0.5}{c} + \frac{0.25}{d},$$

$A^c = \text{"not circle"}$

$$= \frac{1-1}{a} + \frac{1-0.75}{b} + \frac{1-0.5}{c} + \frac{1-0.25}{d} + \frac{1-0}{e}$$

$$= \frac{0.25}{b} + \frac{0.5}{c} + \frac{0.75}{d} + \frac{1}{e},$$

$$B^c = \text{"not square"}$$
$$= \frac{1-0}{a} + \frac{1-0.3}{b} + \frac{1-0.5}{c} + \frac{1-0.7}{d} + \frac{1-1}{e}$$
$$= \frac{1}{a} + \frac{0.7}{b} + \frac{0.5}{c} + \frac{0.3}{d},$$

$$A^c \cap B^c = \text{"neither square nor circle"}$$
$$= \frac{1 \wedge 0}{a} + \frac{0.25 \wedge 0.7}{b} + \frac{0.5 \wedge 0.5}{c} + \frac{0.75 \wedge 0.3}{d} + \frac{1 \wedge 0}{e}$$
$$= \frac{0.25}{b} + \frac{0.5}{c} + \frac{0.3}{d}.$$

1.4.6 Fuzzy Operators

"Not, or, And"; "Extremely, Very, Quite, More, a Little, and Tiny"

Based on the quantitative characterization of the fuzzy concept, we also can define, as certain operations of membership functions, the negative "not", the conjunction "or", "and", and the adverbs "extremely", "very", "quite", "more", "a little", and "tiny" etc. As a result, the fuzzy concept with these words and their derivatives can also be quantified by the membership functions. These defined words and derivatives are called "fuzzy operators."

The operation rules of fuzzy operators are defined as follows:
The membership function of the negative word "not"

$$\mu_{\text{not } \underset{\sim}{A}} = 1 - \mu_{\underset{\sim}{A}};$$

the membership function of the conjunction "or"

$$\mu_{\underset{\sim}{A} \text{ or } \underset{\sim}{B}} = \mu_{\underset{\sim}{A}} \vee \mu_{\underset{\sim}{B}};$$

the membership function of the conjunction "and" ("moreover")

$$\mu_{\underset{\sim}{A} \text{ and } \underset{\sim}{B}} = \mu_{\underset{\sim}{A}} \wedge \mu_{\underset{\sim}{B}}; \left(\mu_{\underset{\sim}{A} \text{ moreover } \underset{\sim}{B}} = \mu_{\underset{\sim}{A}} \wedge \mu_{\underset{\sim}{A}} \right)$$

the membership functions of the adverbs "extremely", "very", "quite", "more", "a little", and "tiny"

$$\mu_{\text{extremely } \underset{\sim}{A}} = \left(\mu_{\underset{\sim}{A}}\right)^4;$$

$$\mu_{\text{very } \underset{\sim}{A}} = \left(\mu_{\underset{\sim}{A}}\right)^2;$$

$$\mu_{\text{quite } \underset{\sim}{A}} = \left(\mu_{\underset{\sim}{A}}\right)^{1.25};$$

$$\mu_{\text{more } \underset{\sim}{A}} = \left(\mu_{\underset{\sim}{A}}\right)^{0.75};$$

$$\mu_{\text{a little } \underset{\sim}{A}} = \left(\mu_{\underset{\sim}{A}}\right)^{0.5};$$

$$\mu_{\text{tiny } \underset{\sim}{A}} = \left(\mu_{\underset{\sim}{A}}\right)^{0.25}.$$

Example 1.13 The membership function value of 0.8 is for a 60-year-old person being the "old" Set. Then the membership function value of $1 - 0.8 = 0.2$ is for those who belong to a new fuzzy set of "not old"; the membership function value of $(0.8)^2 = 0.64$ is for those who are "very old"; the membership function value of $(0.85)^{0.5} = \sqrt{0.8} = 0.9$ is for those who are "a little old", and so on.

Example 1.14 Let the fuzzy sets' membership functions be

$$A_{\sim} = \frac{0.2}{u_1} + \frac{0.5}{u_2} + \frac{0.1}{u_3} + \frac{0.6}{u_4} + \frac{0.8}{u_5},$$

$$B_{\sim} = \frac{0.7}{u_1} + \frac{0.8}{u_2} + \frac{0.4}{u_3} + \frac{0.5}{u_4} + \frac{0.2}{u_5}.$$

Solve: The membership functions of $A_{\sim} \cap B_{\sim}, A_{\sim} \cup B_{\sim}, \overline{A}_{\sim}, \overline{B}_{\sim}$.

Solution:

$$A_{\sim} \cap B_{\sim} = \frac{0.2}{u_1} + \frac{0.5}{u_2} + \frac{0.1}{u_3} + \frac{0.5}{u_4} + \frac{0.8}{u_5};$$

$$A_{\sim} \cup B_{\sim} = \frac{0.7}{u_1} + \frac{0.8}{u_2} + \frac{0.4}{u_3} + \frac{0.6}{u_4} + \frac{0.8}{u_5};$$

$$\overline{A}_{\sim} = \frac{0.8}{u_1} + \frac{0.5}{u_2} + \frac{0.9}{u_3} + \frac{0.4}{u_4} + \frac{0.2}{u_5};$$

$$\overline{B}_{\sim} = \frac{0.3}{u_1} + \frac{0.2}{u_2} + \frac{0.6}{u_3} + \frac{0.5}{u_4} + \frac{0.8}{u_5}.$$

Another example, the degree of a person belonging to the "tall" set is 0.9, and the degree of belonging to the "fat" is only 0.4, then he belongs to the "tall or fat" set with the degree

$$0.9 \vee 0.4 = 0.9;$$

The degree of belonging to the "tall and fat" is only

$$0.9 \wedge 0.4 = 0.4.$$

Example 1.15 Let domain $U = \{1, 2, 3, \ldots, 10\}$. The membership functions of the original words "big" and "small" are

$$\text{"big"} = \frac{0.4}{5} + \frac{0.6}{6} + \frac{0.8}{7} + \frac{1}{8} + \frac{1}{9} + \frac{1}{10};$$
$$\text{"small"} = \frac{1}{1} + \frac{0.8}{2} + \frac{0.6}{3} + \frac{0.2}{4} + \frac{0.1}{5}.$$

then

$$\text{"very big"} = \frac{0.16}{5} + \frac{0.36}{6} + \frac{0.64}{7} + \frac{1}{8} + \frac{1}{9} + \frac{1}{10},$$
$$\text{"extremely big"} = \frac{0.03}{5} + \frac{0.13}{6} + \frac{0.41}{7} + \frac{1}{8} + \frac{1}{9} + \frac{1}{10},$$
$$\text{"very small"} = \frac{1}{1} + \frac{0.64}{2} + \frac{0.36}{3} + \frac{0.04}{4} + \frac{0.01}{5};$$

"neither big nor small" = "not big" $\wedge$ "not small"

$$= \left(\frac{1}{1} + \frac{1}{2} + \frac{1}{3} + \frac{1}{4} + \frac{0.6}{5} + \frac{0.4}{6} + \frac{0.2}{7} \right)$$
$$\wedge \left(\frac{0.2}{2} + \frac{0.4}{3} + \frac{0.8}{4} + \frac{0.9}{5} + \frac{1}{6} + \frac{1}{7} + \frac{1}{8} + \frac{1}{9} + \frac{1}{10} \right)$$
$$= \frac{0.2}{2} + \frac{0.4}{3} + \frac{0.8}{4} + \frac{0.6}{5} + \frac{0.4}{6} + \frac{0.2}{7}.$$

"neither very big nor very small" $= \left(\frac{1}{1} + \frac{1}{2} + \frac{1}{3} + \frac{1}{4} + \frac{0.84}{5} + \frac{0.64}{6} + \frac{0.36}{7} \right)$
$$\wedge \left(\frac{0.36}{2} + \frac{0.64}{3} + \frac{0.96}{4} + \frac{0.99}{5} + \frac{1}{6} + \frac{1}{7} + \frac{1}{8} + \frac{1}{9} + \frac{1}{10} \right)$$
$$= \frac{0.36}{2} + \frac{0.64}{3} + \frac{0.96}{4} + \frac{0.84}{5} + \frac{0.64}{6} + \frac{0.36}{7}.$$

It should be noted that the above do not fully comply with the provisions corresponding to the word meaning of natural language, but they do at best with approximations to the meaning. How to describe more reasonably these adverbs of natural language is worth researching continuously in linguistics.

Example 1.16 If the membership function of the "old" is

$$
\mu_{\text{old}}(x) = \begin{cases} 0, & (x \leq 50), \\ \dfrac{1}{20}(x-50), & (50 < x \leq 70), \\ 1, & (x > 70), \end{cases}
$$

then, the membership function of the "very old" is

$$
\mu_{\text{very old}}(x) = \left[\mu_{\text{old}}(x)^2\right] = \begin{cases} 0, & (x \leq 50), \\ \dfrac{1}{400}(x-50)^2, & (50 < x \leq 70), \\ 1, & (x > 70), \end{cases}
$$

the membership function of the "not very old" is

$$
\mu_{\text{not very old}}(x) = 1 - \mu_{\text{very old}}(x) = \begin{cases} 1, & (x \leq 50), \\ 1 - \dfrac{1}{400}(x-50)^2, & (50 < x \leq 70), \\ 0, & (x > 70). \end{cases}
$$

Notice that it differs from the "very not old", whose membership function is

$$
\mu_{\text{very not old}}(x) = [\mu_{\text{not old}}(x)]^2 = \begin{cases} 1, & (x \leq 50), \\ \left[1 - \dfrac{1}{20}(x-50)\right]^2, & (50 < x \leq 70), \\ 0, & (x > 70). \end{cases}
$$

Thus, by using the membership functions with a small amount of original words, according to certain prior engagement for syntactic rules, we can describe many richer meaning of compound words, which has its own important theoretical significance and practical value.

1.4.7 Cut Sets

In order to classify fuzzy things, the concept of cut sets is introduced, so that fuzzy sets can be converted into common sets.

For example, the "tall" is a fuzzy set, but the "above 1.70 m tall" is a common set because it has a clear line; likewise, the "old" is a fuzzy set, and the "over 70 years old" is a common set. Obviously, with clear boundaries, ambiguity things will be translated into clear things. This threshold is called value λ.

Definition 1.1 In a fuzzy set, the set, made up of the elements whose membership function value is bigger than a certain level value λ, is called cut set λ of fuzzy set, referred to as A_λ, and λ is called the threshold or confidence level, $0 < \lambda \leq 1$.

Obviously, a level set is a common set as is shown in Fig. 1.19.

Fig. 1.19 A level set is a common set

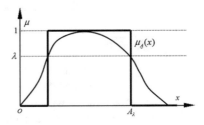

If we write a general expression, namely

$$A_\lambda = \left\{ x \,\middle|\, \mu_{\underset{\sim}{A}}(x) \geq \lambda \right\}, \quad 0 < \lambda \leq 1.$$

Essentially, to take a cut set λ of a fuzzy set $\underset{\sim}{A}$ means to convert the membership function into the characteristic function with the following formula:

$$A_\lambda(x) = \begin{cases} 1, & x \in A_\lambda, \\ 0, & x \notin A_\lambda. \end{cases}$$

Obviously, in fuzzy set $\underset{\sim}{A}$, the elements with their membership degrees bigger than and equal to value λ have made up a common set.

Example 1.17 Let $\underset{\sim}{A} = \dfrac{1}{a} + \dfrac{0.75}{b} + \dfrac{0.5}{c} + \dfrac{0.25}{d} + \dfrac{0}{e}$.

When $\lambda = 1$, $A_1 = \{a\}$.
When $\lambda = 0.6$, $A_{0.6} = \{a, b\}$.
When $\lambda = 0.5$, $A_{0.5} = \{a, b, c\}$.
When $\lambda = 0.2$, $A_{0.2} = \{a, b, c, d\}$.
When $\lambda = 0$, $A_0 = \{a, b, c, d, e\}$.

Obviously, with confidence level λ lower, the more elements of A_λ are expanded with it. In other words, the lower the requirements are, the more the elements (of the fuzzy set) will be qualified too.

Generally, we have a Decomposition Theorem for a practical, effective methods and theory, which can achieve transformation between fuzzy sets and common sets. It shows that a fuzzy set is composed of a common set family.

Exercise 1

1. Let three subsets, $A = \{a, c\}$, $B = \{a, b, c\}$, $C = \emptyset$, of Full Set $U = \{a, b, c, d\}$. Try to write their characteristic functions.

2. Please use keyword technology (such as $f(x) = \dfrac{\text{keywords}}{\text{words of the book name}}$ and alphabetical orders), ordering the following books.

   (1) Keywords: Fuzzy mathematics—by the end of 2013, total 91 books containing the keywords, such as:

      1) Fuzzy mathematics, the bush tax policy guide;
      2) Fuzzy mathematics and engineering sciences;
      3) Fuzzy mathematical methods in economic management;
      4) Fuzzy mathematical methods in risk researches of inter-current diseases —Fuzzy clustering analysis;
      5) Fuzzy mathematics to comprehensive evaluation of user satisfaction;
      6) Application of fuzzy mathematics in artificial intelligence;
      7) Application of fuzzy mathematics in chemistry;
      8) Application of fuzzy mathematics in ship engineering;
      9) Application of fuzzy mathematics in construction management.

   (2) Keywords: fuzzy, application—by the end of 2013, total 510 books containing these two keywords, such as:

      1) Fuzzy sets' methods and their application in meteorology;
      2) Application of fuzzy mathematics in assets share pricing of life insurance;
      3) Application of fuzzy systems and improved ANFIS in space optics;
      4) Dynamic fuzzy neural network: design and application;
      5) Fuzzy economy: Surviving in internet economic times;
      6) Using fuzzy neural inference system—ANFIS forecast for world bulk fleet capacity;
      7) English vocabulary fuzzy shorthand;
      8) Introduction to TCM fuzzy Methods;
      9) Fuzzy sets' analyses in psychology;
      10) On the ideological and political work of fuzzy intermediaries;
      11) Fuzzy system and fuzzy control tutorial;
      12) Fuzzy portfolio optimization: Theory and methods;
      13) Fuzzy seismology;
      14) Fuzzy set theory and management decision-making;
      15) Underground water quality assessment based on fuzzy decision theory;
      16) Fuzzy linguistics;
      17) Fuzzy neural network;
      18) Fuzzy prediction;
      19) Stochastic programming and fuzzy programming;

20) Adaptive fuzzy control theory and its application;

21) Fuzzy integral model in traffic environment quality.

3. Let fuzzy sets A, B be

$$A = \frac{0.1}{x_1} + \frac{0.2}{x_2} + \frac{0.7}{x_3} + \frac{0.8}{x_4} + \frac{0.9}{x_5};$$

$$B = \frac{0.7}{x_1} + \frac{0.5}{x_2} + \frac{0.3}{x_3} + \frac{0.2}{x_4} + \frac{0.4}{x_5}.$$

Solve: $A \cup B, A \cap B, \overline{A}, \overline{B}$.

4. Let domain $U = \{u_1, u_2, u_3, u_4, u_5\}$, $A\sim = \{0.5, 0.1, 0, 1, 0.8\}$, $B = \{0.1, 0.4, 0.9, 0.7, 0.2\}$, $C = \{0.8, 0.2, 1, 0.4, 0.3\}$. Calculate $A \cup B, A \cap B$, $\left(A \cup B \right) \cap C, \overline{A}$.

5. In domain $U = \{u_1, u_2, u_3, u_4, u_5\}$, let its fuzzy subsets be:

$$A = \frac{0.8}{u_1} + \frac{0.4}{u_2} + \frac{0.5}{u_3} + \frac{1}{u_4} + \frac{0.3}{u_5};$$

$$B = \frac{0.6}{u_1} + \frac{0.4}{u_2} + \frac{0.8}{u_4} + \frac{0.2}{u_5};$$

$$C = \frac{0.5}{u_1} + \frac{0.4}{u_2} + \frac{0.7}{u_4} + \frac{0.3}{u_5};$$

$$D = \frac{0.4}{u_1} + \frac{0.6}{u_2} + \frac{0.2}{u_4} + \frac{0.8}{u_5}.$$

Try to determine if the following relations are correct:

(1) $B \subseteq A$;   (2) $C \subseteq B$;   (3) $D = \overline{B}$.

6. Take domain $U = \{1, 2, 3, 4, 5, 6, 7, 8, 9, 10\}$, "a small number" is denoted by A, but "close to the number 10" is denoted by B, try to write the expressions of A and B in U.

7. Let domain $U = R$ (real number field), A is "the real number larger than 5". Try to write the membership function of A.

8. Let the membership function of "young" be

$$\mu_A(x) = \begin{cases} 1, & 0 \le x \le 25, \\ \left[1 + \left(\frac{x-25}{5} \right)^2 \right]^{-1} & 25 < x \le 100. \end{cases}$$

(1) Make the membership function curves;

(2) Solve $\mu_A(30), \mu_A(40), \mu_A(45)$;

(3) Write the membership function of "very young", and draw the curves;

(4) For $x = 30, 40, 45$, solve the membership degrees of "very young".

9. Let the membership function of "old" be

$$\mu_{\underset{\sim}{A}}(x) = \begin{pmatrix} 0, & 0 \le x \le 50, \\ \left[1 + \left(\frac{x-25}{5}\right)^2\right]^{-1} & 50 < x \le 100. \end{pmatrix}$$

(1) Make the membership function curves;

(2) Solve the membership functions of $\overline{\underset{\sim}{A}}$ in the last Exercise No. 8 and $\overline{\underset{\sim}{B}}$ of this No. 9;

(3) For $x = 30, 40, 45$, solve the membership degrees of $\overline{\underset{\sim}{B}}, \overline{\underset{\sim}{A}}, \overline{\underset{\sim}{B}}$ respectively.

10. Let fuzzy set $\underset{\sim}{A} = \frac{0.2}{a} + \frac{0}{b} + \frac{0.1}{c} + \frac{0.6}{d} + \frac{1}{e} + \frac{0.4}{f} + \frac{0.8}{g}$. Solve the Cut Sets $A_{0.3}$, $A_{0.6}, A_{0.8}$.

Summary

I. The Knowledge Structures of This Chapter

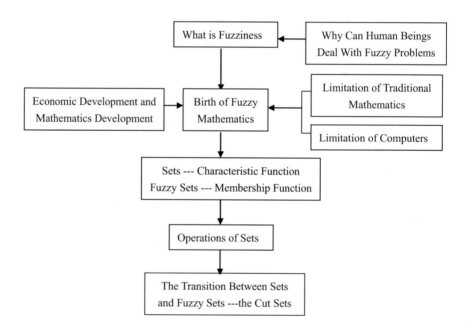

II. Review and Reflection

1. Excessive accurateness makes things worse, but proper fuzziness turns better.
2. Fuzzy mathematics is bound to be born with needs of economic development and limitations of traditional mathematics and computer experiences.
3. Recalling Sets representations, extending from characteristic functions to membership functions, and representing fuzzy sets by membership functions.
4. Recalling the subsets, complement, intersection, combination of the sets, leading to the operations of fuzzy sets.
5. The transition between sets and fuzzy sets—the cut sets.

Chapter 2
Matrix, Fuzzy Relation and Fuzzy Matrix

2.1 Introduction

2.1.1 Matrix Represents a Form; Matrix Describes a Network

A matrix is a common mathematical tool, a way to record and manage a large volume of data. It can store information succinctly. By matrix operations, you can easily process into information.

Example 2.1 In a supermarket for convenience of consumers, it serves dishes to be sent to the doors by reservation. In Table 2.1, it describes the served dishes on Monday, and the households' reserved number.

Table 2.1 The dishes and the number reserved by households

Dishes \ Households	A	B	C	D
Zhao	2	1	0	0
Li	0	1	2	1
Wang	1	1	2	2

Table 2.2 The digital table

$$A = \begin{pmatrix} 2 & 1 & 0 & 0 \\ 0 & 1 & 2 & 1 \\ 1 & 1 & 2 & 2 \end{pmatrix}$$

In fact, we need not make the table, as detailed as that in Table 2.1. If we are dealing with such a table every day, we can only record the number in the matrix (see Table 2.2). We can prearrange the meaning of each number in the table. The horizontal lines are set as rows, and the vertical lines are set as columns. For instance, the

© Springer International Publishing AG, part of Springer Nature 2018
H.-R. Lin et al., *Fuzzy Sets Theory Preliminary*,
https://doi.org/10.1007/978-3-319-70749-5_2

third column shows the quantity of Dish C reserved by each household. The second line indicates that the customer named Li reserves a variety of dishes. This rectangular arrangement is the matrix, often expressed in uppercase letters, such as A, B and so on. We call the digital Table 2.2 a matrix of 3 rows and 4 columns.

In addition, a matrix can show a table, and a matrix can also be used to describe the network. The so-called network is the graphics composed by the nodes and links. For example, if A, B, C are referred to three villages, there are highways between them, as is shown in Fig. 2.1. The villages are regarded as nodes, and the highways as links, forming a network. If the villages are linked with one direct highway between them, it is denoted by 1; if the villages are linked without any direct highway between them, denoted by 0; if the villages are linked with two direct highways between them, denoted by 2 (for example, 1 is recorded to AC; 2 is recorded to AB).

Example 2.2 Figure 2.1 shows that the connectivity of a network can be described by the following matrix:

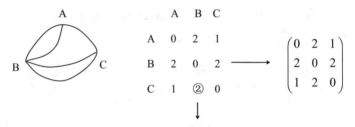

Fig. 2.1 The graph of a connected network

There are 2 connecting lines between the third row C and the second column B, and so on.

Practice 1 Try to use a matrix to denote the following network diagram.

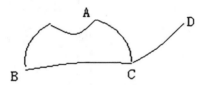

2.1.2   A Matrix Can Also be Used to Describe the Relation

Example 2.3 For two men games, let $X = \{$stone, scissors, cloth$\}$, the "win" as 1, the "draw" as 0.5, the "loss" as 0, then the "Relation between A and B" can be expressed by a matrix.

$$R = \begin{array}{c} \\ \text{stone} \\ \text{scissors} \\ \text{cloth} \end{array} \begin{array}{ccc} \text{stone} & \text{scissors} & \text{cloth} \\ \begin{pmatrix} 0.5 & 1 & 0 \\ 0 & 0.5 & 1 \\ 1 & 0 & 0.5 \end{pmatrix} \end{array}.$$

2.2 The Concept of a Matrix

Definition 2.1 The data table made up of m × n of the number a_{ij} ($i = 1, 2, \ldots m$; $j = 1, 2, \ldots n$)

$$\begin{pmatrix} a_{11} & a_{12} & \cdots & a_{1n} \\ a_{21} & a_{22} & \cdots & a_{2n} \\ \cdots & \cdots & \cdots & \cdots \\ a_{m1} & a_{m2} & \cdots & a_{mn} \end{pmatrix},$$

is called a matrix of m × n which is arranged in m rows and n columns; the number of m × n is called elements; the number a_{ij} ($i = 1, 2, \ldots m; j = 1, 2, \ldots n$) is called an element in Row i and Column j. A matrix can be represented by the capital letters, such as A, B, C, or, (a_{ij}), (b_{ij}) ... In order to indicate the number of columns and rows of a matrix, sometimes it is recorded as $A_{m \times n}$ or $(a_{ij})_{m \times n}$.

Practice 2 If the matrix elements of the row and column $= \begin{pmatrix} 1 & 2 & 7 \\ 6 & -3 & 8 \\ 4 & 5 & 0 \end{pmatrix}$ are referred to as $a_{ij}(i = 1, 2, 3)$, $a_{ij}(j = 1, 2, 3)$, then $a_{23} = \underline{\hspace{1cm}}$, $a_{32} = \underline{\hspace{1cm}}$, $a_{22} = \underline{\hspace{1cm}}$.

If $A = (a_{ij})$, $B = (b_{ij})$ are all a matrix of $m \times n$, and their corresponding elements are equal, i.e., $a_{ij} = b_{ij}(i = 1, 2, \ldots, m, \ j = 1, 2, \ldots, n)$, then matrix A is equal to matrix B, denoted as $A = B$.

Here are some special matrices.

(1) When all the elements are zero, the matrix is a zero matrix, denoted by 0, namely

$$0 = \begin{pmatrix} 0 & 0 & \cdots & 0 \\ 0 & 0 & \cdots & 0 \\ \cdots & \cdots & \cdots & \cdots \\ 0 & 0 & \cdots & 0 \end{pmatrix}.$$

(2) When $m = 1$, the matrix has only one row

$$A = (a_{11}, a_{12}, \ldots, a_{1n}),$$

A is called a row matrix.

(3) When $n = 1$, the matrix has only one column

$$A = \begin{pmatrix} a_{11} \\ a_{21} \\ \cdots \\ a_{m1} \end{pmatrix},$$

A is called a column matrix.

(4) When $m = n$, the

$$A = \begin{pmatrix} a_{11} & a_{12} & \cdots & a_{1n} \\ a_{21} & a_{22} & \cdots & a_{2n} \\ \cdots & \cdots & \cdots & \cdots \\ a_{n1} & a_{n2} & \cdots & a_{nn} \end{pmatrix},$$

A is called an n-order square.

(5) In the following form,

$$E_n = \begin{pmatrix} 1 & 0 & \cdots & 0 \\ 0 & 1 & \cdots & 0 \\ \cdots & \cdots & \cdots & \cdots \\ 0 & 0 & \cdots & 1 \end{pmatrix}$$

the diagonal elements are all 1, and the remaining elements are all an n-order square of 0, called an n-order unit matrix.

2.3   Matrix's Addition, Subtraction and Multiplication

We discuss the matrix operation by way of examples.

As Example 2.1, if we need to know the total dishes of each customer's order on Monday and Tuesday, we must add up each corresponding number of the two matrices together on Monday and Tuesday. Thus the method combining two matrices into a third matrix is called matrix addition, for example,

$$\begin{pmatrix} 2 & 1 & 0 & 0 \\ 0 & 1 & 2 & 1 \\ 1 & 1 & 2 & 2 \end{pmatrix} + \begin{pmatrix} 2 & 1 & 2 & 1 \\ 2 & 0 & 0 & 1 \\ 0 & 2 & 0 & 2 \end{pmatrix} = \begin{pmatrix} 4 & 2 & 2 & 1 \\ 2 & 1 & 2 & 2 \\ 1 & 3 & 2 & 4 \end{pmatrix}.$$

It is not difficult to see that two matrices can be added up only when the orders of them are the same, i.e., only when they have the same "shape" of the matrices.

2.3.1 Matrix's Addition and Subtraction

Definition 2.2 Suppose there are two matrices of $m \times n$

$$A = \begin{pmatrix} a_{11} & a_{12} & \cdots & a_{1n} \\ a_{21} & a_{22} & \cdots & a_{2n} \\ \cdots & \cdots & \cdots & \cdots \\ a_{m1} & a_{m2} & \cdots & a_{mn} \end{pmatrix}, B = \begin{pmatrix} b_{11} & b_{12} & \cdots & b_{1n} \\ b_{21} & b_{22} & \cdots & b_{2n} \\ \cdots & \cdots & \cdots & \cdots \\ b_{m1} & b_{m2} & \cdots & b_{mn} \end{pmatrix},$$

then the sum of matrix A and B is recorded as $A + B$, and is defined as:

$$A + B = \begin{pmatrix} a_{11} + b_{11} & a_{12} + b_{12} & \cdots & a_{1n} + b_{1n} \\ a_{21} + b_{21} & a_{22} + b_{21} & \cdots & a_{2n} + b_{2n} \\ \cdots & \cdots & \cdots & \cdots \\ a_{m1} + b_{m1} & a_{m2} + b_{m2} & \cdots & a_{mn} + b_{mn} \end{pmatrix}.$$

Only the same number of rows and columns of two matrices can be added up.
The matrix

$$\begin{pmatrix} -a_{11} & -a_{12} & \cdots & -a_{1n} \\ -a_{21} & -a_{22} & \cdots & -a_{2n} \\ \cdots & \cdots & \cdots & \cdots \\ -a_{m1} & -a_{m2} & \cdots & -a_{mn} \end{pmatrix}$$

is called matrix A's $A = \begin{pmatrix} a_{11} & a_{12} & \cdots & a_{1n} \\ a_{21} & a_{22} & \cdots & a_{2n} \\ \cdots & \cdots & \cdots & \cdots \\ a_{m1} & a_{m2} & \cdots & a_{mn} \end{pmatrix}.$

Negative matrix is denoted as $-A$.

Therefore, the above matrix can be used to define the subtraction of matrices, namely $A - B = A + (-B)$.

The operation of matrix addition satisfies the following rules:

(1) $A + B = B + A$ (Commutative law);
(2) $(A + B) + C = A + (B + C)$ (Associative law);

(3) $A + 0 = 0 + A = A$;

(4) $A + (-A) = (-A) + A = 0$.

Try to prove the above yourself.

2.3.2 Multiplication of a Number and a Matrix

As Example 2.1, a supermarket serves dish reservation, if the dishes reserved on Tuesday are the same as those on Monday, the total dishes for two days are

$$\begin{pmatrix} 2 & 1 & 0 & 0 \\ 0 & 1 & 2 & 1 \\ 1 & 1 & 2 & 2 \end{pmatrix} \times 2 = \begin{pmatrix} 4 & 2 & 0 & 0 \\ 0 & 2 & 4 & 2 \\ 2 & 2 & 4 & 4 \end{pmatrix}.$$

We can see that the matrix multiplied by the number is, in fact, that each element of the matrix is multiplied by this number.

Definition 2.3 Let K be a number, A is a matrix of $m \times n$

$$A = \begin{pmatrix} a_{11} & a_{12} & \cdots & a_{1n} \\ a_{21} & a_{22} & \cdots & a_{2n} \\ \cdots & \cdots & \cdots & \cdots \\ a_{m1} & a_{m2} & \cdots & a_{mn} \end{pmatrix}.$$

Then the product of K and matrix A, referred to as KA or AK, is defined as

$$KA = AK = \begin{pmatrix} ka_{11} & ka_{12} & \cdots & ka_{1n} \\ ka_{21} & ka_{22} & \cdots & ka_{2n} \\ \cdots & \cdots & \cdots & \cdots \\ ka_{m1} & ka_{m2} & \cdots & ka_{mn} \end{pmatrix}.$$

The product of a number and a matrix is that all the elements of this matrix are multiplied by the number.

The multiplication of a number and a matrix satisfies the following operation rules (where K, L is numbers, A, B are matrices of $m \times n$):

(1) $(K + L) A = KA + LA$, (Distributive law)

   $K (A + B) = KA + KB$.

(2) $(KL) A = K (LA) = L (KA)$, (Associative law)

(3) $1 \cdot A = A, -1 \cdot A = -A$.

(4) If $KA = 0$, then $K = 0$ or $A = 0$.

Please complete the proof all by yourselves.

Example 2.4 Suppose to $A = \begin{pmatrix} 1 & 0 & 1 \\ 2 & -1 & 3 \end{pmatrix}$, $B = \begin{pmatrix} 2 & -1 & 0 \\ 3 & 2 & 5 \end{pmatrix}$, then

$$2A = \begin{pmatrix} 2 & 0 & 2 \\ 4 & -2 & 6 \end{pmatrix}, \quad 3B = \begin{pmatrix} -6 & 3 & 0 \\ -9 & -6 & -15 \end{pmatrix},$$

$$2A - 3B = \begin{pmatrix} -4 & 3 & 2 \\ -5 & -8 & -9 \end{pmatrix}.$$

2.3.3 Multiplication of Matrices

As Example 2.1, a supermarket serves dish reservation, if Dish A costs 2 yuan, B costs 3 yuan, C 4 yuan, and D 5 yuan, how much should each household pay for the dishes on Monday?

Take the customer, Mr. Zhao for an example. What he has reserved is shown in the first row of the table, such as the row matrix $(2 \quad 1 \quad 0 \quad 0)$. We put the prices of the dishes in a column matrix $\begin{pmatrix} 2 \\ 3 \\ 4 \\ 5 \end{pmatrix}$, which is then so-called "multiplied by" the row matrix $(2 \quad 1 \quad 0 \quad 0)$.

$$(2 \quad 1 \quad 0 \quad 0) \begin{pmatrix} 2 \\ 3 \\ 4 \\ 5 \end{pmatrix} = 2 \times 2 + 1 \times 3 + 0 \times 4 + 0 \times 5 = 4 + 3 = 7 \text{ Yuan}.$$

Here the "multiplication" is actually that each element of the row matrix is multiplied by each corresponding element of the column matrix, and then added up.

If three clients: Zhao, Li, and Wang are taken into account, the matrix of reservation can be "multiplied" by the matrix of prices:

$$\begin{pmatrix} 2 & 1 & 0 & 0 \\ 0 & 1 & 2 & 1 \\ 1 & 1 & 2 & 2 \end{pmatrix} \begin{pmatrix} 2 \\ 3 \\ 4 \\ 5 \end{pmatrix} = \begin{pmatrix} 2 \times 2 + 1 \times 3 + 0 \times 4 + 0 \times 5 \\ 0 \times 2 + 1 \times 3 + 2 \times 4 + 1 \times 5 \\ 1 \times 2 + 1 \times 3 + 2 \times 4 + 2 \times 5 \end{pmatrix} = \begin{pmatrix} 7 \\ 16 \\ 23 \end{pmatrix}.$$

If on Tuesday the prices have been adjusted into Dish A at 2.1 yuan, B at 3.6 yuan, C at 4 yuan, D at 6 yuan, then the calculation can be made as follows:

$$
\begin{array}{c}
\quad \begin{array}{cc} \text{Monday} & \text{Tuesday} \\ \text{Price} & \text{Price} \end{array}
\end{array}
$$

$$
\begin{array}{c} \text{Zhao} \\ \text{Li} \\ \text{Wang} \end{array}
\begin{pmatrix} 2 & 1 & 0 & 0 \\ 0 & 1 & 2 & 1 \\ 1 & 1 & 2 & 2 \end{pmatrix}
\begin{pmatrix} 2 & 2.1 \\ 3 & 3.6 \\ 4 & 4 \\ 5 & 6 \end{pmatrix}
$$

$$
= \begin{pmatrix}
2\times2+1\times3+0\times4+0\times5 & 2\times2.1+1\times3.6+0\times4+0\times6 \\
0\times2+1\times3+2\times4+1\times5 & 0\times2.1+1\times3.6+2\times4+1\times6 \\
1\times2+1\times3+2\times4+2\times5 & 1\times2.1+1\times3.6+2\times4+2\times6
\end{pmatrix}
$$

$$
= \begin{pmatrix} 7 & 7.8 \\ 16 & 17.6 \\ 23 & 25.7 \end{pmatrix} \begin{array}{l} \text{Zhao} \\ \text{Li} \\ \text{Wang} \end{array}
$$

$$
\begin{array}{cc} \text{Monday} & \text{Tuesday} \\ \text{Total Payment} & \text{Total Payment} \end{array}
$$

It is not difficult to see from the above example that when the multiplication of two matrices happens, we always multiply each row of the first matrix by each column of the second matrix. Therefore, only when the row elements of the first matrix are equal to the column elements of the second matrix, this multiplication can be made. If the above two numbers are unequal, then they can not be multiplied.

Definition 2.4 Let $A = (a_{ij})$ be a matrix of m × s, $B = (b_{ij})$ is a matrix of $s \times n$, namely

$$
A = \begin{pmatrix} a_{11} & a_{12} & \cdots & a_{1s} \\ a_{21} & a_{22} & \cdots & a_{2s} \\ & \cdots & \cdots & \\ a_{m1} & a_{m2} & \cdots & a_{ms} \end{pmatrix}, B = \begin{pmatrix} b_{11} & b_{12} & \cdots & b_{1n} \\ b_{21} & b_{22} & \cdots & b_{2n} \\ & \cdots & \cdots & \\ b_{s1} & b_{s2} & \cdots & b_{sn} \end{pmatrix}.
$$

Then the product of matrix A and B is matrix $C = (C_{ij})$ of $m \times n$, in which

$$
C_{ij} = a_{i1}b_{1j} + a_{i2}b_{j2} + \cdots + a_{is}b_{sj} = \sum_{k-1}^{s} a_{ik}b_{kj},
$$

$$
(i = 1, 2, \ldots, m; j = 1, 2, \ldots, n).
$$

And this product is referred to as $C = AB$.

They can be multiplied by each other only when the number of columns in the left matrix A is equal to that of rows in the right matrix B, and when the number of rows of product C is equal to that of A, and the number of columns of C is equal to that of B.

Example 2.5 Let the $A = \begin{pmatrix} 2 & -1 & -1 \\ 3 & 1 & 2 \end{pmatrix}, B = \begin{pmatrix} 1 & 3 \\ 2 & -1 \\ -5 & 0 \end{pmatrix}$, Then

$$AB = \begin{pmatrix} 2 & -1 & -1 \\ 3 & 1 & -2 \end{pmatrix} \begin{pmatrix} 1 & 3 \\ 2 & -1 \\ -5 & 0 \end{pmatrix}$$

$$= \begin{pmatrix} 2.1 + (-1) \cdot 2 + (-1) \cdot (-5) & 2.3 + (-1) \cdot (-1) + (-1) \cdot 0 \\ 3.1 + 1.2 + (-2) \cdot (-5) & 3.3 + 1 \cdot (-1) + (-2) \cdot 0 \end{pmatrix}$$

$$= \begin{pmatrix} 5 & 7 \\ 15 & 8 \end{pmatrix},$$

and

$$BA = \begin{pmatrix} 1 & 3 \\ 2 & -1 \\ -5 & 0 \end{pmatrix} \begin{pmatrix} 2 & -1 & -1 \\ 3 & 1 & -2 \end{pmatrix}$$

$$= \begin{pmatrix} 1 \times 2 + 3 \times 3 & 1 \times (-1) + 3 \times 1 & 1 \times (-1) + 3 \times (-2) \\ 2 \times 2 + (-1) \times 3 & 2 \times (-1) + (-1) \times 1 & 2 \times (-1) + (-1) \times (-2) \\ (-5) \times 2 + 0 \times 3 & (-5) \times (-1) + 0 \times 1 & (-5) \times (-1) + 0 \times (-2) \end{pmatrix}$$

$$= \begin{pmatrix} 11 & 2 & -7 \\ 1 & -3 & 0 \\ -10 & 5 & 5 \end{pmatrix}.$$

Example 2.6 Let $A = \begin{pmatrix} 2 & 4 \\ -3 & -6 \end{pmatrix}, B = \begin{pmatrix} -2 & 4 \\ 1 & -2 \end{pmatrix}$. Then

$$AB = \begin{pmatrix} 2 & 4 \\ -3 & -6 \end{pmatrix} \begin{pmatrix} -2 & 4 \\ 1 & -2 \end{pmatrix} = \begin{pmatrix} 0 & 0 \\ 0 & 0 \end{pmatrix},$$

$$BA = \begin{pmatrix} -2 & 4 \\ 1 & -2 \end{pmatrix} \begin{pmatrix} 2 & 4 \\ -3 & -6 \end{pmatrix} = \begin{pmatrix} -16 & -32 \\ 8 & 16 \end{pmatrix}.$$

Example 2.7 Let $A = \begin{pmatrix} 1 & 1 \\ -1 & -1 \end{pmatrix}, B = \begin{pmatrix} 1 & -1 \\ -1 & 1 \end{pmatrix}, C = \begin{pmatrix} 3 & -3 \\ -3 & 3 \end{pmatrix}$. Then

$$AB = \begin{pmatrix} 1 & 1 \\ -1 & -1 \end{pmatrix}\begin{pmatrix} 1 & -1 \\ -1 & 1 \end{pmatrix} = \begin{pmatrix} 0 & 0 \\ 0 & 0 \end{pmatrix},$$

$$BA = \begin{pmatrix} 1 & -1 \\ -1 & 1 \end{pmatrix}\begin{pmatrix} 1 & 1 \\ -1 & -1 \end{pmatrix} = \begin{pmatrix} 2 & 2 \\ -2 & -2 \end{pmatrix},$$

$$AC = \begin{pmatrix} 1 & 1 \\ -1 & -1 \end{pmatrix}\begin{pmatrix} 3 & -3 \\ -3 & 3 \end{pmatrix} = \begin{pmatrix} 0 & 0 \\ 0 & 0 \end{pmatrix}.$$

From the above examples, we know

(1) The matrix multiplication can not satisfy the commutative law, i.e. in general, $AB \neq BA$;
(2) The product of two non-zero matrices may be zero matrix, i.e. when $AB = 0$, it can not be $A = 0$ or $B = 0$ generally. Therefore, the matrix multiplication can not satisfy cancellation law, that is, if $AB = AC$, $A \neq 0$, it can not be $B = C$, i.e. generally A can not be eliminated from both sides of the equation.

If matrixes A and B satisfy $AB = BA$, then A and B are commutative.

Example 2.8 Given the following matrix

$$A = \begin{pmatrix} 2 & 7 & 9 \\ -3 & 1 & -5 \end{pmatrix}, B = \begin{pmatrix} 3 & -1 \\ 4 & 0 \\ -2 & 6 \end{pmatrix}, C = \begin{pmatrix} -6 & 4 \\ 1 & 11 \\ 0 & -3 \end{pmatrix}.$$

According to the results, calculate and guess what operation laws the matrix multiplication can satisfy.

(1) $A (B + C)$ and $AB + AC$;
(2) $(B + C) A$ and $BA + CA$;
(3) $K (AB)$ and $A (KB)$ ($K \in R$).

Solution

(1) $B + C = \begin{pmatrix} 3 & -1 \\ 4 & 0 \\ -2 & 6 \end{pmatrix} + \begin{pmatrix} -6 & 4 \\ 1 & 11 \\ 0 & -3 \end{pmatrix} = \begin{pmatrix} -3 & 3 \\ 5 & 11 \\ -2 & 3 \end{pmatrix},$

$$A(B + C) = \begin{pmatrix} 2 & 7 & 9 \\ -3 & 1 & -5 \end{pmatrix}\begin{pmatrix} -3 & 3 \\ 5 & 11 \\ -2 & 3 \end{pmatrix} = \begin{pmatrix} 11 & 110 \\ 24 & -13 \end{pmatrix},$$

$$AB + AC = \begin{pmatrix} 2 & 7 & 9 \\ -3 & 1 & -5 \end{pmatrix} \begin{pmatrix} 3 & -1 \\ 4 & 0 \\ -2 & 6 \end{pmatrix} + \begin{pmatrix} 2 & 7 & 9 \\ -3 & 1 & -5 \end{pmatrix} \begin{pmatrix} -6 & 4 \\ 1 & 11 \\ 0 & -3 \end{pmatrix}$$

$$= \begin{pmatrix} 16 & 52 \\ 5 & -27 \end{pmatrix} + \begin{pmatrix} -5 & 58 \\ 19 & 14 \end{pmatrix} = \begin{pmatrix} 11 & 110 \\ 24 & -13 \end{pmatrix}.$$

(2) $(B + C)A = \begin{pmatrix} -3 & 3 \\ 5 & 11 \\ -2 & 3 \end{pmatrix} \begin{pmatrix} 2 & 7 & 9 \\ -3 & 1 & -5 \end{pmatrix} = \begin{pmatrix} -15 & -18 & -42 \\ -23 & 46 & -10 \\ -13 & -11 & -33 \end{pmatrix},$

$$BA + CA = \begin{pmatrix} 3 & -1 \\ 4 & 0 \\ -2 & 6 \end{pmatrix} \begin{pmatrix} 2 & 7 & 9 \\ -3 & 1 & -5 \end{pmatrix} + \begin{pmatrix} -6 & 4 \\ 1 & 11 \\ 0 & -3 \end{pmatrix} \begin{pmatrix} 2 & 7 & 9 \\ -3 & 1 & -5 \end{pmatrix}$$

$$= \begin{pmatrix} 9 & 20 & 32 \\ 8 & 28 & 36 \\ -22 & -8 & -48 \end{pmatrix} + \begin{pmatrix} -24 & -38 & -74 \\ -31 & 18 & -46 \\ 9 & -3 & 15 \end{pmatrix}$$

$$= \begin{pmatrix} -15 & -18 & -42 \\ -23 & 46 & -10 \\ -13 & -11 & -33 \end{pmatrix}.$$

(3) $K(AB) = K \begin{pmatrix} 16 & 52 \\ 5 & -27 \end{pmatrix} = \begin{pmatrix} 16K & 25K \\ 5K & -27K \end{pmatrix},$

$$A(KB) = \begin{pmatrix} 2 & 7 & 9 \\ -3 & 1 & -5 \end{pmatrix} \begin{pmatrix} 3K & -K \\ 4K & 0 \\ -2K & 6K \end{pmatrix} = \begin{pmatrix} 16K & 25K \\ 5K & -27K \end{pmatrix}.$$

From the above calculation, the matrix multiplication can be guessed (when matrix multiplication is meaningful) to satisfy the following operational laws:

① The distributive law of matrix multiplication to addition:
$A(B + C) = AB + AC$ and $(B + C)A = BA + CA$.
② The associative law of matrix multiplication with real numbers:
$K(AB) = (KA)B = A(KB)$ (K is an arbitrary number).

2.4   Relation

Here we describe how to represent fuzzy relation by using fuzzy matrices, and how
to deal with the fuzzy relation by fuzzy matrix operations.

2.4.1   The Concept of Relation

In the objective world, all kinds of things have relation universally. One of the
mathematical models describing the links among them is the relation, commonly
represented by the symbol: R.

Example 2.9 Suppose that there exist two Sets of X and Y, where $X = \{x \mid x$ is the
table tennis team players of Class A$\}$ and $Y = \{y \mid y$ is that of Class B$\}$. Class A has
three table tennis players, recorded as 1, 2, 3; Class B has three table tennis players,
recorded as a, b, c. If R represents the rival relation between Class A and Class B,
then it is recorded as *1Ra* that a match is arranged between 1 and a; it is denoted as
2Rb that a match is made between 2 and b; so it is with *3Rc* etc.

If Element X and Y have no relation with each other, it is represented by $\bar{R}$. As
there is no match between 1 and b, 2 and c in the above example, it can be referred
to as $1 \bar{R}$ b and $2 \bar{R}$ c respectively, and so on.

If R represents the common relation from set Y to Set X, then for any
$x \in X, y \in Y$, it can only have one of the following two conditions:

1. x and y have some relation R, namely xRy;
2. x and y have no relation R, namely $x\bar{R}y$.

The relation R from X to Y can also be represented by the ordered pairs (x, y),
such that $x \in X, y \in Y$, all the ordered pairs can make set R.

Definition 2.5 Choose an element from set X and Set Y respectively, and arrange
them in ordered pairs. All such pairs constitute a new set called Cartesian product
set of X and Y, denoted by

$$X \times Y = \{(x, y) \mid x \in X, y \in Y\}.$$

Apparently, set R is a subset of Cartesian product set of X and Y, namely
$R \subset X \times Y$.

As the above examples, Cartesian product set of X and Y

$$X \times Y = \{(1, a), (1, b), (1, c), (2, a), (2, b), (2, c), (3, a), (3, b), (3, c)\},$$

and

$$R = \{(1, a), (2, b), (3, c)\},$$

Apparently

$$R \subset X \times Y.$$

Generally, the relation from A to B differs from that from B to A.

Definition 2.6 If R is defined as the relation from X to Y, and Y equals X, then R can be said to be the relation of set X.

Example 2.10 Suppose that $A = \{1, 2, 3, 4\}$, if R represents the relation of "a is smaller than b" in set A, then

$$R = \{(1, 2), (1, 3), (1, 4), (2, 3), (2, 4), (3, 4)\}.$$

R in each ordered pair of the first element and the second element has a "smaller than" relation.

Example 2.11 The Cartesian product of $X \times Y$ such that point set $X = \{x\}$ in axis X and point set $Y = \{y\}$ in axis Y is a set of all points (x, y) of the coordinate plane. This relation of "Abscissa is greater than ordinate" is a set of Part I and Part III quadrant's points below the angular bisector, shown in Fig. 2.2.

Fig. 2.2 The relation of "x > y"

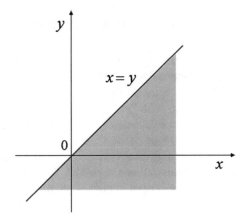

2.4.2 Relation Representation

For the relation of two sets of X and Y, in general, if Set X has m elements and Set Y has n elements, then we can use Matrix R to represent the relation from set X to set Y.

$$R = \begin{pmatrix} r_{11} & r_{12} & \cdots & r_{1n} \\ r_{21} & r_{22} & \cdots & r_{2n} \\ \cdots & \cdots & \cdots & \cdots \\ r_{m1} & r_{m2} & \cdots & r_{mn} \end{pmatrix},$$

where $r_{ij} = 0$ or 1, $i = 1, 2, \ldots, m$; $j = 1, 2, \ldots, n$.

If $r_{ij} = 1$, it represents that the ith element of set X has a relation R with the jth element of set Y; if $r_{ij} = 0$, it represents that the ith element of set X has no relation R with the jth element of set Y.

As Example 2.9, the relation of table tennis matches can be expressed as follows

$$A = \begin{matrix} & \begin{matrix} a & b & c \end{matrix} \\ \begin{matrix} 1 \\ 2 \\ 3 \end{matrix} & \begin{pmatrix} 1 & 0 & 0 \\ 0 & 1 & 0 \\ 0 & 0 & 1 \end{pmatrix} \end{matrix}.$$

As Example 2.10, the relation of "a is smaller than b" can be expressed as follows

$$B = \begin{matrix} & \begin{matrix} 1 & 2 & 3 & 4 \end{matrix} \\ \begin{matrix} 1 \\ 2 \\ 3 \\ 4 \end{matrix} & \begin{pmatrix} 0 & 1 & 1 & 1 \\ 0 & 0 & 1 & 1 \\ 0 & 0 & 0 & 1 \\ 0 & 0 & 0 & 0 \end{pmatrix} \end{matrix}.$$

2.5 Fuzzy Relation and Fuzzy Matrix

A fuzzy relation is extension of an ordinary relation, playing an important role in fuzzy mathematics.

In an ordinary set theory, Relation R describes the relation between affirmation and negation with "yes" and "no". It has a range of $\{0, 1\}$. 1 is yes, 0 is no.

Besides the "absolutely" relation with "yes" and "no", the objective things also have many vague concepts, such as "some relation", "close relation" and so on. For example, "Moderate water has a great relation with crop yield.", "A resemblance exists between father and child.", "Fertilization has some relation with crop pests.", "Price changes have important relation with people's lives.", "RMB exchange rate has correlation with foreign trade," etc. The relation like the above have no clear boundary which can only be portrayed by means of fuzzy relation.

When we indicate the fuzzy relation by matrix, the a_{ij} should be able to represent the membership fuzzy relation $\underset{\sim}{R}$ of the ith element in Set X, partly related to the jth element of set Y, i.e., $\mu_{\underset{\sim}{R}}(x, y)$ also reflecting an extension of the relation between X and Y.

Definition 2.7 Let X and Y be any two sets. In Cartesian product set

$$X \times Y = \{(x, y) | x \in X, y \in Y\},$$

fuzzy subset R is called fuzzy relation of X and Y, its membership function is denoted by $\mu_R(x, y) \in [0, 1]$. When $X = Y$, fuzzy subset R of $X \times X$ is called fuzzy relation on X.

Obviously, fuzzy relation is a special kind of fuzzy set.

Example 2.12 Let $X = \{x_1, x_2, x_3\}$, represent a set of three people: A_1, B_1, C_1; $Y = \{y_1, y_2\}$ represent a set of two people: A, B.

$$R = \frac{0.4}{(x_1, y_1)} + \frac{0.8}{(x_1, y_2)} + \frac{0.9}{(x_2, y_1)} + \frac{0.1}{(x_2, y_2)} + \frac{0.6}{(x_3, y_1)} + \frac{0.9}{(x_3, y_2)}$$

determines a "familiar" fuzzy relation on $X \times Y$. For example, $R(x_2, y_2) = 0.1$ interprets that B_1 and B are not so familiar; $R(x_3, y_2) = 0.9$ represents that C_1 and B are very familiar. Such "familiar" relation R on $X \times Y$ can be expressed in Table 2.3.

Table 2.3 The relation of "familiar"

R	A	B
A1	0.4	0.8
B1	0.9	0.1
C1	0.6	0.9

Of course, this table can also be simply referred to as a matrix form:

$$\begin{pmatrix} 0.4 & 0.8 \\ 0.9 & 0.1 \\ 0.6 & 0.9 \end{pmatrix}.$$

Definition 2.8 We define the matrix as fuzzy matrix, whose elements contain values within the closed interval [0, 1].

When X and Y are finite domains, fuzzy relation R of X to Y can be represented by fuzzy matrix.

In general, $X = \{x_1, x_2, \ldots, x_m\}$ and $Y = \{y_1, y_2, \ldots, y_n\}$ are finite sets, the fuzzy relation of $X \times Y$ can be represented by the following $m \times n$ matrix:

$$\begin{pmatrix} \mu_R(x_1, y_1) & \mu_R(x_1, y_2) & \cdots & \mu_R(x_1, y_n) \\ \mu_R(x_2, y_1) & \mu_R(x_2, y_2) & \cdots & \mu_R(x_2, y_n) \\ \cdots & \cdots & \cdots & \cdots \\ \mu_R(x_m, y_1) & \mu_R(x_m, y_2) & \cdots & \mu_R(x_m, y_n) \end{pmatrix}$$

Obviously, when $X = Y$, $\underset{\sim}{R}$ is the fuzzy relation on X, if $X = \{x_1, x_2, \ldots, x_n\}$, then $\underset{\sim}{R}$ can be represented as a matrix of n orders.

The general form of fuzzy matrix is

$$A = \begin{pmatrix} a_{11} & a_{12} & \cdots & a_{1n} \\ a_{21} & a_{22} & \cdots & a_{2n} \\ \cdots & \cdots & \cdots & \cdots \\ a_{m1} & a_{m2} & \cdots & a_{mn} \end{pmatrix},$$

where $0 \leq a_{ij} \leq 1, 1 \leq i \leq m, 1 \leq j \leq n$, it is recorded as $\underset{\sim}{A} = \left[a_{ij}\right]_{m \times n}$.

Example 2.13 We examine the relation $\underset{\sim}{R}$ of "much bigger than" on set $X = \{1, 5, 7, 9, 20\}$. Cartesian product set of $X \times X$ has twenty ordered pairs altogether. The first element is indeed much bigger than the second one in the ordered pair $(20, 1)$. It could be considered that the membership degree for the "much more than" would be 1. The first element is bigger than the second element in the ordered pair $(9, 5)$, but it is not much bigger. It can be considered that the membership degree of it for the "much more than" would be 0.3. Apparently, it is quite subjective to determine the value of $\underset{\sim}{R}$'s membership function like this, but it is also an objective reflection of fuzzy relation to X. By similarly discussing, we can determine

$$\underset{\sim}{R} = \frac{0.5}{(5,1)} + \frac{0.7}{(7,1)} + \frac{0.1}{(7,5)} + \frac{0.8}{(9,1)} + \frac{0.3}{(9,5)} + \frac{0.1}{(9,7)}$$
$$+ \frac{1}{(20,1)} + \frac{0.95}{(20,5)} + \frac{0.90}{(20,7)} + \frac{0.85}{(20,9)}.$$

It can also be expressed as

$$\underset{\sim}{R} = \begin{array}{c} 1 \\ 5 \\ 7 \\ 9 \\ 20 \end{array} \begin{pmatrix} \begin{array}{ccccc} 1 & 5 & 7 & 9 & 20 \end{array} \\ \begin{array}{ccccc} 0 & 0 & 0 & 0 & 0 \\ 0.5 & 0 & 0 & 0 & 0 \\ 0.7 & 0.1 & 0 & 0 & 0 \\ 0.8 & 0.3 & 0.1 & 0 & 0 \\ 1 & 0.95 & 0.90 & 0.85 & 0 \end{array} \end{pmatrix}$$

In general, as long as it is given to the membership function $\mu_{\underset{\sim}{R}}(x, y)$ of fuzzy set $\underset{\sim}{R}$ in Cartesian product set of $X \times Y$, the fuzzy relation $\underset{\sim}{R}$ from set X to set Y is also determined.

Example 2.14 Medically the people's standard weight is expressed by: weight (kg) = height (cm) − 100, which actually gives the binary relation of the height (X) and weight (Y). Let $X = \{140, 150, 160, 170, 180\}$, $Y = \{40, 50, 60, 70, 80\}$, then the above relation can be expressed as

$$R = \begin{array}{c} \\ 140 \\ 150 \\ 160 \\ 170 \\ 180 \end{array} \begin{array}{ccccc} 40 & 50 & 60 & 70 & 80 \\ \left(\begin{array}{ccccc} 1 & 0 & 0 & 0 & 0 \\ 0 & 1 & 0 & 0 & 0 \\ 0 & 0 & 1 & 0 & 0 \\ 0 & 0 & 0 & 1 & 0 \\ 0 & 0 & 0 & 0 & 1 \end{array} \right) \end{array}$$

People are fat or thin differently, for a "non-standard" situation, it should be described by the extent of its closeness to their standard. So the fuzzy relation expressed by the following fuzzy matrix can reflect apparently fuller the relation between height and standard weight.

$$\underset{\sim}{R} = \begin{array}{c} \\ 140 \\ 150 \\ 160 \\ 170 \\ 180 \end{array} \begin{array}{ccccc} 40 & 50 & 60 & 70 & 80 \\ \left(\begin{array}{ccccc} 1 & 0.8 & 0.2 & 0.1 & 0 \\ 0.8 & 1 & 0.8 & 0.2 & 0.1 \\ 0.2 & 0.8 & 1 & 0.8 & 0.2 \\ 0.1 & 0.2 & 0.8 & 1 & 0.8 \\ 0 & 0.1 & 0.2 & 0.8 & 1 \end{array} \right) \end{array}.$$

Example 2.15 Let $X = Y$ for the Set of real numbers. The fuzzy relation $\underset{\sim}{R}$ of "x_1 far greater than x_2" is defined as:

$$\underset{\sim}{R}(x_1, x_2) = \begin{cases} 0, & x_1 \le x_2, \\ \left[1 + \dfrac{100}{(x_1 - x_2)^2} \right]^{-1}, & x_1 > x_2. \end{cases}$$

For example, $(1, 0) \approx 0.001$ indicates that the degree of "1 far greater than 0" is only 0.001, while $(101, 1) \approx 0.909$ indicates that the degree of "101 is much larger than 1" is 0.909. Here the real numbers set is an infinite set.

2.6 Operations of Fuzzy Relation

A fuzzy relation is a fuzzy subset on $X \times Y$, being a special kind of fuzzy sets. So the arithmetic nature of fuzzy sets in Chap. 1 is the same as that of fuzzy relation.

The following are five common operations:

Suppose $\underset{\sim}{R_1}$ and $\underset{\sim}{R_2}$ are fuzzy relations on $X \times Y, \forall (x, y) \in X \times Y$, then there are:

(1) Equality:

$$\underset{\sim}{R_1} = \underset{\sim}{R_2} \Leftrightarrow \mu_{\underset{\sim}{R_1}}(x, y) = \mu_{\underset{\sim}{R_2}}(x, y).$$

(2) Containing:

$$R_1 \subseteq R_2 \Leftrightarrow \mu_{R_1}(x,y) \le \mu_{R_2}(x,y).$$

(3) Union:

$$R_1 \cup R_2 \Leftrightarrow \mu_{(R_1 \cup R_2)}(x,y) = \text{Max}\{\mu_{R_1}(x,y), \mu_{R_2}(x,y)\}$$
$$= \mu_{R_1}(x,y) \vee \mu_{R_2}(x,y).$$

(4) Intersection:

$$R_1 \cap R_2 \Leftrightarrow \mu_{(R_1 \cap R_2)}(x,y) = \text{Min}\{\mu_{R_1}(x,y), \mu_{R_2}(x,y)\}$$
$$= \mu_{R_1}(x,y) \wedge \mu_{R_2}(x,y).$$

(5) Complement:

$$\overline{R}_1 \Leftrightarrow \mu_{\overline{R}_1}(x,y) = 1 - \mu_{R_1}(x,y).$$

Example 2.16 Suppose R_1, R_2, R_3 and R_4 are fuzzy relations on $X \times Y$, $(x,y) \in X \times Y$, and

$$R_1 = \begin{pmatrix} 0.5 & 0.4 \\ 0.7 & 0.6 \end{pmatrix} \begin{matrix} y_1 \\ y_2 \end{matrix} \quad R_2 = \begin{pmatrix} 0.7 & 0.1 \\ 0.2 & 0.9 \end{pmatrix} \begin{matrix} y_1 \\ y_2 \end{matrix},$$

$$R_3 = \begin{pmatrix} \frac{1}{2} & \frac{2}{5} \\ \frac{7}{10} & \frac{3}{5} \end{pmatrix} \begin{matrix} y_1 \\ y_2 \end{matrix}, \quad R_4 = \begin{pmatrix} 0.8 & 0.8 \\ 0.9 & 0.6 \end{pmatrix} \begin{matrix} y_1 \\ y_2 \end{matrix}.$$

Try to find the relation of equality and containing, and seek $R_1 \cup R_2, R_1 \cap R_4, R_2 \cap R_4, \overline{R}_1$.

Obviously, $R_1 = R_3, R_1 \subseteq R_4, R_3 \subseteq R_4,$

$$R_1 \cup R_2 = \begin{pmatrix} 0.5 & 0.4 \\ 0.7 & 0.6 \end{pmatrix} \vee \begin{pmatrix} 0.7 & 0.1 \\ 0.2 & 0.9 \end{pmatrix}$$
$$= \begin{pmatrix} 0.5 \vee 0.7 & 0.4 \vee 0.1 \\ 0.7 \vee 0.2 & 0.6 \vee 0.9 \end{pmatrix}$$
$$= \begin{pmatrix} 0.7 & 0.4 \\ 0.7 & 0.9 \end{pmatrix}.$$

$$R_1 \cap R_4 = \begin{pmatrix} 0.5 & 0.4 \\ 0.7 & 0.6 \end{pmatrix} \wedge \begin{pmatrix} 0.8 & 0.8 \\ 0.9 & 0.6 \end{pmatrix}$$

$$= \begin{pmatrix} 0.5 \wedge 0.8 & 0.4 \wedge 0.8 \\ 0.7 \wedge 0.9 & 0.6 \wedge 0.6 \end{pmatrix}$$

$$= \begin{pmatrix} 0.5 & 0.4 \\ 0.7 & 0.6 \end{pmatrix} = R_1 \quad (\because R_1 \subseteq R_4).$$

$$R_2 \cap R_4 = \begin{pmatrix} 0.7 & 0.1 \\ 0.2 & 0.9 \end{pmatrix} \wedge \begin{pmatrix} 0.8 & 0.8 \\ 0.9 & 0.6 \end{pmatrix}$$

$$= \begin{pmatrix} 0.7 \wedge 0.8 & 0.1 \wedge 0.8 \\ 0.2 \wedge 0.9 & 0.9 \wedge 0.6 \end{pmatrix}$$

$$= \begin{pmatrix} 0.7 & 0.1 \\ 0.2 & 0.6 \end{pmatrix}.$$

$$\because \quad \mu_{\overline{R_1}}(x, y) = 1 - \mu_{R_1}(x, y),$$

$$\therefore \quad \overline{R_1} = \begin{pmatrix} 1 - 0.5 & 1 - 0.4 \\ 1 - 0.7 & 1 - 0.6 \end{pmatrix} = \begin{pmatrix} 0.5 & 0.6 \\ 0.3 & 0.4 \end{pmatrix}.$$

2.7   Synthesis of Fuzzy Relation

2.7.1   Origin of Synthesis

In ordinary relation, two relations can be synthesized into a new relation, such as

$$\left. \begin{array}{l} a \text{ is the mother of } b \\ b \text{ is the father of } c \end{array} \right\} \quad a \text{ is the grandmother of } c.$$

In the fuzzy relation, the similar situation happens too, such as

$$\left. \begin{array}{l} \text{The age of } a \text{ is similar to } b \\ b \text{ is much older than } c \end{array} \right\} \quad a \text{ is much older than } c \text{ too.}$$

It is also a complex, called Synthesis in fuzzy relation.

2.7.2 Definition and Operations of Synthesis

Definition 2.9 Let R_1 be a fuzzy relation from X to Y, and R_2 be a fuzzy relation from Y to Z. Then the synthetic relation $R = R_1 \circ R_2$ is fuzzy synthetic relation from X to Z. Its membership function is

$$\mu_R(x, z) = \mu_{(R_1 \circ R_2)}(x, z)$$
$$= \text{MaxMin}[\mu_{R_1}(x, y), \mu_{R_1}(y, z)]$$
$$= \vee[\mu_{R_1}(x, y) \wedge \mu_{R_2}(y, z)].$$

Example 2.17 Let

$$R_1 = \begin{pmatrix} x_1 & x_2 \\ 1 & 0.4 \\ 0.2 & 0.5 \end{pmatrix} \begin{matrix} y_1 \\ y_2, \end{matrix} \qquad R_2 = \begin{pmatrix} y \\ 0.5 \\ 0.6 \end{pmatrix} \begin{matrix} z_1 \\ z_2. \end{matrix}$$

Then

$$\mu_R(x, z) = \mu_{(R_1 \circ R_2)}(x, z)$$
$$= \begin{pmatrix} 1 & 0.4 \\ 0.2 & 0.5 \end{pmatrix} \circ \begin{pmatrix} 0.5 \\ 0.6 \end{pmatrix}.$$

The operations of synthetic relation is similar to matrix multiplication, but the multiplication sign must be changed into the "$\wedge$" operator, and the plus sign into the "$\vee$" operator in the matrix multiplication. Then the above formula will be

$$R = \begin{pmatrix} (1 \wedge 0.5) \vee (0.4 \wedge 0.6) \\ (0.2 \wedge 0.5) \vee (0.5 \wedge 0.6) \end{pmatrix}$$
$$= \begin{pmatrix} 0.5 \vee 0.4 \\ 0.2 \vee 0.5 \end{pmatrix}$$
$$= \begin{pmatrix} 0.5 \\ 0.5 \end{pmatrix}.$$

Example 2.18 Suppose that $A = \begin{pmatrix} 0.3 & 0.2 & 0.1 \\ 0.2 & 0.4 & 0.6 \end{pmatrix}$, $B = \begin{pmatrix} 0.1 & 0.2 \\ 0.5 & 0.3 \\ 0.6 & 0.4 \end{pmatrix}$, then

$$A \circ B = \begin{pmatrix} (0.3 \wedge 0.1) \vee (0.2 \vee 0.5) \vee (0.1 \wedge 0.6) & (0.3 \wedge 0.2) \vee (0.2 \wedge 0.3) \vee (0.1 \wedge 0.4) \\ (0.2 \wedge 0.1) \vee (0.4 \wedge 0.5) \vee (0.6 \wedge 0.6) & (0.2 \wedge 0.2) \vee (0.4 \wedge 0.3) \vee (0.6 \wedge 0.4) \end{pmatrix}$$

$$= \begin{pmatrix} 0.2 & 0.2 \\ 0.6 & 0.4 \end{pmatrix}.$$

Similarly we can solve

$$B \circ A = \begin{pmatrix} 0.2 & 0.2 & 0.2 \\ 0.3 & 0.3 & 0.3 \\ 0.3 & 0.4 & 0.4 \end{pmatrix}.$$

So $A \circ B \neq B \circ A$.

Then a few more practical examples will illustrate the significance of synthesis.

Example 2.19 Suppose that the similar relation R that children look similar to their parents in a family is a fuzzy relation and can be expressed as shown in Table 2.4,

Table 2.4 The similar relation between children and parents

$\mu_R(x,y)$	Father	Mother
Son	0.8	0.2
Daughter	0.1	0.6

that is

$$R = \begin{pmatrix} 0.8 & 0.2 \\ 0.1 & 0.6 \end{pmatrix}.$$

There is also another similar relation that parents look similar to their grandparents, which is expressed as shown in Table 2.5,

Table 2.5 The similar relation between parents and grandparents

$\mu_S(y,z)$	Grandfather	Grandmother
Father	0	0.1
Mother	0.5	0.7

that is

$$S = \begin{pmatrix} 0 & 0.1 \\ 0.5 & 0.7 \end{pmatrix},$$

then $\mu_{(R \circ S)}(x,z)$ represents that the similar relation that children look similar to their grandparents, i.e.,

$$R \circ S = \begin{pmatrix} 0.8 & 0.2 \\ 0.1 & 0.6 \end{pmatrix} \circ \begin{pmatrix} 0 & 0.1 \\ 0.5 & 0.7 \end{pmatrix}$$

$$= \begin{pmatrix} (0.8 \wedge 0) \vee (0.2 \wedge 0.5) & (0.8 \wedge 0.1) \vee (0.2 \wedge 0.7) \\ (0.1 \wedge 0) \vee (0.6 \wedge 0.5) & (0.1 \wedge 0.1) \vee (0.6 \wedge 0.7) \end{pmatrix}$$

$$= \begin{pmatrix} 0 \vee 0.2 & 0.1 \vee 0.2 \\ 0 \vee 0.5 & 0.1 \vee 0.6 \end{pmatrix}$$

$$= \begin{pmatrix} 0.2 & 0.2 \\ 0.5 & 0.6 \end{pmatrix}.$$

So children and grandparents looking similar to each other are expressed as shown in Table 2.6:

Table 2.6 The similar relation between children and grandparents	$\mu_{(R \circ S)}(x, z)$	Grandfather	Grandmother
	Son	0.2	0.2
	Daughter	0.5	0.6

In the above statements, $X = \{son, daughter\}$, $Y = \{father, mother\}$; $Z = \{grandfather, grandmother\}$, the result of synthesis is that the daughter looks more like her grandmother, which is popularly called "three generations would look like each other".

Example 2.20 Let R_{12} represent the relation between the color and maturity of tomatoes, and R_2 represent the relation between the maturity and taste of tomatoes. Suppose that the color domain $X = \{green, yellow, red\}$, the maturity levels domain $Y = \{immature, half-mature, mature\}$, and the taste domain $Z = \{sour, tasteless, sweet\}$, where

$$R_1 = \begin{array}{c} \\ green \\ yellow \\ red \end{array} \begin{array}{ccc} immature & half\text{-}mature & mature \\ \begin{pmatrix} 1 & 0.2 & 0 \\ 0.3 & 1 & 0.4 \\ 0 & 0.2 & 1 \end{pmatrix} \end{array},$$

$$R_2 = \begin{array}{c} \\ immature \\ half\text{-}mature \\ mature \end{array} \begin{array}{ccc} sour & tasteless & sweet \\ \begin{pmatrix} 1 & 0.2 & 0 \\ 0.7 & 1 & 0.3 \\ 0 & 0.7 & 1 \end{pmatrix} \end{array}.$$

In accordance with synthetic operations of fuzzy relation, the synthetic relation between colors and tastes of tomatoes is calculated as following

$$\underset{\sim}{R}_1 \circ \underset{\sim}{R}_2 = \begin{array}{c} \\ green \\ yellow \\ red \end{array} \begin{pmatrix} \begin{array}{ccc} sour & tasteless & sweet \\ 1 & 0.5 & 0.15 \\ 0.7 & 1 & 0.4 \\ 0.14 & 0.7 & 1 \end{array} \end{pmatrix}.$$

Example 2.21 To see a doctor of (TCM) traditional Chinese medicine, let $X = \{x_1, x_2, x_3\}$ be a set of patients. A set of symptoms $Y = \{y_1 \text{ (cough)}, y_2 \text{ (fever)}, y_3 \text{ (sweat)}, y_4 \text{ (cold)}\}$. Diseases set $Z = \{z_1 \text{ (having a cold)}, z_2 \text{ (pneumonia)}\}$. The existing fuzzy relation between patients and symptoms $\underset{\sim}{R} : X \rightarrow Y$. The fuzzy relation between symptoms and diseases $\underset{\sim}{R} : Y \rightarrow Z$. According to the diagnosis experience of famous TCM doctors, the degrees of symptoms related with diseases are shown in Table 2.7.

Table 2.7 The degrees of symptoms related with diseases	$\underset{\sim}{S}$	Z_1	Z_2
	Y_1	0.6	0.4
	Y_2	0.4	0.7
	y_3	0.8	0.6
	y_4	0.8	0.4

There are three patients. x_1 says he has a little cough with tepid palms and abnormal sweat. x_2 says he has a bad cough with a little fever and sweat, but feels cold. x_3 says he has little cough, but has a fever and fatigue with much sweat a little cold. By talking and diagnosis of patients, the degrees of symptoms are shown in Table 2.8.

Table 2.8 The patients' talking and diagnosis and degrees of symptoms	$\underset{\sim}{R}$	y_1	y_2	y_3	y_4
	x_1	0.3	0.6	0.4	0
	x_2	0.7	0.4	0.1	0.6
	x_3	0.1	0.9	0.6	0.3

What would be wrong with them by diagnosis?

Solution From

$$
\underset{\sim}{R} \circ \underset{\sim}{S} = \begin{pmatrix} 0.3 & 0.6 & 0.4 & 0 \\ 0.7 & 0.4 & 0.1 & 0.6 \\ 0.1 & 0.9 & 0.6 & 0.3 \end{pmatrix} \circ \begin{pmatrix} 0.6 & 0.4 \\ 0.4 & 0.7 \\ 0.8 & 0.6 \\ 0.8 & 0.4 \end{pmatrix}
$$

$$
= \begin{pmatrix} (0.3 \wedge 0.6\} \vee (0.6 \wedge 0.4) \vee (0.4 \wedge 0.8) \vee (0 \wedge 0.8) \\ (0.7 \wedge 0.6) \vee (0.4 \wedge 0.4) \vee (0.1 \wedge 0.8) \vee (0.6 \wedge 0.8) \\ (0.1 \wedge 0.6) \vee (0.9 \wedge 0.4) \vee (0.6 \wedge 0.8) \vee (0.3 \wedge 0.8) \end{pmatrix}
$$
$$
\begin{pmatrix} (0.3 \wedge 0.4) \vee (0.6 \wedge 0.7) \vee (0.4 \wedge 0.6) \vee (0 \wedge 0.4) \\ (0.7 \wedge 0.4) \vee (0.4 \wedge 0.7) \vee (0.1 \wedge 0.6) \vee (0.6 \wedge 0.4) \\ (0.1 \wedge 0.4) \vee (0.9 \wedge 0.7) \vee (0.6 \wedge 0.6) \vee (0.3 \wedge 0.4) \end{pmatrix}
$$

$$
= \begin{pmatrix} 0.3 \vee 0.4 \vee 0.4 \vee 0 & 0.3 \vee 0.6 \vee 0.4 \vee 0 \\ 0.6 \vee 0.4 \vee 0.1 \vee 0.6 & 0.4 \vee 0.4 \vee 0.1 \vee 0.4 \\ 0.1 \vee 0.4 \vee 0.6 \vee 0.3 & 0.1 \vee 0.7 \vee 0.6 \vee 0.3 \end{pmatrix}
$$

$$
= \begin{pmatrix} 0.4 & 0.6 \\ 0.6 & 0.4 \\ 0.6 & 0.7 \end{pmatrix},
$$

we can get (Table 2.9).

Table 2.9 The results from diagnosis	$R \circ S$	z_1	z_2
	x_1	0.4	<u>0.6</u>
	x_2	<u>0.6</u>	0.4
	x_3	0.6	<u>0.7</u>

i.e., Patient x_1 and Patient x_3 are truly possible to suffer from pneumonia. Patient x_2 are rather possible to suffer from a cold.

A University of Traditional Chinese Medicine has used these principles to develop a computer software, whose tested diagnoses are similar to experts'.

Obviously, fuzzy matrix serves as a representation of fuzzy relation on a finite domain, so the operations of fuzzy relation are also available to fuzzy matrix.

The relevant concepts of fuzzy matrix are introduced again as follows.

1. The Power of Fuzzy Matrix

Definition 2.10 Let $\underset{\sim}{A}$ be an n-order fuzzy matrix. So

$$\underset{\sim}{A^1} = \underset{\sim}{A},$$

$$\underset{\sim}{A^2} = \underset{\sim}{A} \circ \underset{\sim}{A},$$

$$\underset{\sim}{A^3} = \underset{\sim}{A^2} \circ \underset{\sim}{A},$$

$$\cdots$$

$$\underset{\sim}{A^k} = \underset{\sim}{A^{k-1}} \circ \underset{\sim}{A}.$$

Then, $\underset{\sim}{A^k}$ is called the k-th power of $\underset{\sim}{A}$.

2. Transpose of a Fuzzy Matrix

Definition 2.11 Transpose from each other of the rows and columns of a fuzzy matrix $\underset{\sim}{A} = (a_{ij})_{m \times n}$ to a new fuzzy matrix $(a_{ji})_{n \times m}$, which is called a transpose matrix, represented by $\underset{\sim}{A^T} = (a_{ji})_{n \times m}$.

Transpose of a fuzzy matrix corresponds to converse relation of corresponding fuzzy relation.

3. A Fuzzy Symmetric Square Matrix

Definition 2.12 Let $\underset{\sim}{A}$ be a fuzzy square matrix, and $\underset{\sim}{A} = \underset{\sim}{A^T}$. Then $\underset{\sim}{A}$ is a fuzzy symmetric square matrix.

Example 2.22 Given $\underset{\sim}{R} = \begin{pmatrix} 0.2 & 0.5 \\ 0.4 & 0.9 \end{pmatrix}$, then

$$
\begin{aligned}
\underset{\sim}{R^2} = \underset{\sim}{R} \circ \underset{\sim}{R} &= \begin{pmatrix} 0.2 & 0.5 \\ 0.4 & 0.9 \end{pmatrix} \circ \begin{pmatrix} 0.2 & 0.5 \\ 0.4 & 0.9 \end{pmatrix} \\
&= \begin{pmatrix} (0.2 \wedge 0.2) \vee (0.5 \wedge 0.4) & (0.2 \wedge 0.5) \vee (0.5 \wedge 0.9) \\ (0.4 \wedge 0.2) \vee (0.9 \wedge 0.4) & (0.4 \wedge 0.5) \vee (0.9 \wedge 0.9) \end{pmatrix} \\
&= \begin{pmatrix} 0.4 & 0.5 \\ 0.4 & 0.9 \end{pmatrix}.
\end{aligned}
$$

$$\underset{\sim}{R^3} = \underset{\sim}{R^2} \circ \underset{\sim}{R} = \begin{pmatrix} 0.4 & 0.5 \\ 0.4 & 0.9 \end{pmatrix} \circ \begin{pmatrix} 0.2 & 0.5 \\ 0.4 & 0.9 \end{pmatrix}$$

$$= \begin{pmatrix} 0.2 \vee 0.4 & 0.4 \vee 0.5 \\ 0.2 \vee 0.4 & 0.4 \vee 0.9 \end{pmatrix}$$

$$= \begin{pmatrix} 0.4 & 0.5 \\ 0.4 & 0.9 \end{pmatrix}.$$

Exercise 2

1. The gasoline sales (in liters) of the two gas stations are given by the following matrix:

Gasoline With Four-star Brand   Gasoline With Two-star Brand

$$\begin{matrix} A \\ B \end{matrix} \begin{pmatrix} 300 & 100 \\ 200 & 200 \end{pmatrix}$$

(1) The price of gasoline with four-star brand is CNY10 per liter, two-star brand of gasoline CNY9 per liter, write the data in a 2×1 matrix;

(2) Multiply one matrix by the other to get a 2×1 matrix $\begin{pmatrix} x \\ y \end{pmatrix}$.

(3) Write down the value of $x + y$. What information does this number contain?

(4) One yuan profit per liter is made on the four-star brand, and two yuan profits per liter on the two-star brand. Multiply the two matrices to obtain profits of the gas station.

2. Draw the network described by the following matrix:

$$\text{from} \quad \begin{array}{c} \\ A \\ B \\ C \\ D \end{array} \overset{\displaystyle \overset{\text{to}}{A \quad B \quad C \quad D}}{\begin{bmatrix} 0 & 1 & 2 & 1 \\ 1 & 0 & 1 & 1 \\ 2 & 1 & 0 & 1 \\ 1 & 1 & 1 & 0 \end{bmatrix}}.$$

3. Draw and complete the following diagram to illustrate the set $\{2, 3, 4, 6, 8\}$ by the relation of "is a quality factor of ...", write a matrix to represent this relation.

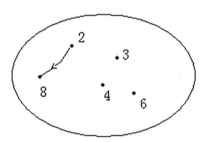

4. The following figure shows the relation of "... is a sister of ..." in the set of four children {a, b, c, d}.

   (1) Which of the children are boys?
   (2) Draw a diagram showing the relationship of "... regards ... as a brother" in the same set;
   (3) Write a matrix to demonstrate the relationship of "... regards ... as a brother".

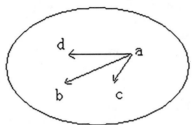

5. Given that matrix $A = \begin{pmatrix} 2 & -3 \\ 4 & -5 \end{pmatrix}$, matrix $B = \begin{pmatrix} -4 & 2 \\ -3 & 7 \end{pmatrix}$, according to the following conditions, solve $C = \lambda A + \mu B$, where $\lambda = 2, \mu = 1$.

6. If possible, calculate the product of the following.

   (1) $(3 \quad 2)\begin{pmatrix} -5 \\ 4 \end{pmatrix}$;   (2) $(2 \quad -3 \quad 7)\begin{pmatrix} 4 \\ 2 \\ -5 \end{pmatrix}$;

   (3) $(8 \quad 0 \quad 5)\begin{pmatrix} 2 \\ 7 \end{pmatrix}$;   (4) $(6 \quad -5 \quad 7 \quad 0)\begin{pmatrix} -1 \\ 4 \\ 3 \\ 9 \end{pmatrix}$.

7. If $A = \begin{pmatrix} 1 & 0 & 3 \\ 2 & -1 & 0 \end{pmatrix}, B = \begin{pmatrix} 1 & -1 \\ 2 & 3 \\ 4 & 0 \end{pmatrix}$, try to solve AB and BA.

8. $A = \begin{pmatrix} 1 & 0 \\ 0 & 2 \end{pmatrix}, B = \begin{pmatrix} -1 & 1 \\ 0 & -2 \end{pmatrix}, C = \begin{pmatrix} 2 & -3 \\ -1 & 0 \end{pmatrix}$.

   (1) Calculate $A\,(BC)$;
   (2) Calculate $(AB)\,C$;
   (3) What nature do the above results illustrate for matrix multiplication?

9. If $x = \begin{pmatrix} 1 & 0 & 2 \\ -3 & 2 & 0 \\ -1 & 1 & 3 \end{pmatrix}, y = \begin{pmatrix} -3 & -2 & 2 \\ -9 & 7 & 3 \\ 2 & 1 & 0 \end{pmatrix}, z = \begin{pmatrix} 2 & 0 \\ 0 & -1 \\ 3 & 2 \end{pmatrix}$, if possible, try to calculate the following formulas.

   (1) $x - y$;   (2) $x + z$;   (3) xy;   (4) xz;   (5) y^2;   (6) yxz.

10. Solve the values of p and q to establish the equality $\begin{pmatrix} p & 5 \\ 4 & q \end{pmatrix} \begin{pmatrix} 2 \\ 1 \end{pmatrix} = \begin{pmatrix} 7 \\ 3 \end{pmatrix}$.

11. $A = \begin{pmatrix} 1 & 0 \\ 2 & -1 \\ -3 & 2 \end{pmatrix}$, write a transpose of A, take it as A^T, calculate: (1) AA^T;

(2) $A^T A$.

12. If $A = \begin{pmatrix} 4 & 4 \\ -2 & -1 \end{pmatrix}$, $B = \begin{pmatrix} 2 & 1 \\ -2 & 1 \end{pmatrix}$, calculate:

(1) $A^2 - B^2$;   (2) $(A + B)(A - B)$.
Explain why $A^2 - B^2 \neq (A + B)(A - B)$.

13. Calculate:

(1) $\begin{pmatrix} a_1 & 0 & 0 \\ 0 & a_2 & 0 \\ 0 & 0 & a_3 \end{pmatrix}^5$;   (2) $\begin{pmatrix} 0 & 1 & 0 \\ 0 & 0 & 1 \\ 0 & 0 & 0 \end{pmatrix}^3$;

(3) $\begin{pmatrix} \cos\theta & \sin\theta \\ -\sin\theta & \cos\theta \end{pmatrix}^k$, k is a positive integer.

14. Airline A has some airplanes, such as 8 Boeing, 6 McDonnell Douglas and 2 Airbus aircrafts;
Airline B has 9 Boeing, 1 McDonnell Douglas and 7 Airbus aircrafts;
Airline C has 2 Boeing, 11 McDonnell Douglas aircrafts.

(1) Demonstrate the data with matrix A of 3×3;
(2) Every Boeing aircraft can carry 180 passengers, every McDonnell Douglas aircraft 200 passengers, and every Airbus aircraft 230 passengers. Write two appropriate matrices so that their product can determine the number of passengers at full load for each airline;
(3) Solve the product of these two matrices.

15. The investors A, B, and C purchased the following shares in a stock market at the same time on April 1, 2004. The quantity and names of the purchased shares are shown separately in the following table.

Investors	Shares' names				
	Taiji industry	Chunlan shares	Yuanshui shares	ST-tongda	Celebrity electricity
	Quantity (shares)				
A	1000	0	1000	0	1000
B	500	500	800	400	300
C	2000	1000	3000	0	0

Then the shares' prices are listed as follows: Taiji industry 4.37 yuan/share; Chunlan shares 7.08 yuan/share; Yuanshui shares 8.12 yuan/share; ST-tongda 15.20 yuan/share;
Celebrity electricity 13.45 yuan/share.

(1) Use a matrix to demonstrate the shares' quantity bought by each person;
(2) Calculate the time invested in the stock market funds per matrix;
(3) If the current closing price of each stock is listed as follows: Taiji industry 4.45 yuan/share; Chunlan shares 6.72 yuan/share; Yuanshui shares 8.22 yuan/share; ST-tongda 13.98 yuan/share; Celebrity electricity 12.56 yuan/share, calculate the current incomes for each person by a matrix (excluding transaction fees).

16. A grocery store can make three different kinds of birthday cakes: Type A, B, and C according to the customers' booking. Each cake's ingredient proportion and the price per kilogram (CNY) are shown in the following table (the ingredients weight in kilograms):

	Fruit	Oil	Sugar	Flour	Egg	Wine
A	0.3	0.6	0.8	0.1	0.6	0.3
B	0.2	0.4	0.6	0.1	0.4	0.2
C	0.1	0.2	0.4	0.1	0.2	0.1
Ingredient Price (CNY/kg)	6	5	4	1	5	4

One day, according to the customers' booking, the grocery store has made 4 cakes with Type A, 6 cakes with B, and 8 cakes with C. Try to solve the total costs of all the booked cakes and the cost of each cake with three types that day.

17. Suppose that $R_1 = \begin{pmatrix} 0.1 & 0 & 0.8 \\ 0.9 & 0.5 & 0 \\ 0 & 0.4 & 0.3 \end{pmatrix}, R_2 = \begin{pmatrix} 0.7 & 0.2 & 0.4 \\ 0.3 & 0.1 & 0.6 \\ 1 & 0.5 & 0.2 \end{pmatrix},$

solve: $R_1 \cup R_2, \overline{(R_1 \cap R_2)}, \overline{R_1}.$

18. Suppose that $R_1 = \begin{pmatrix} 0.7 & 0.4 & 0.1 & 1 \\ 0.8 & 0.3 & 0.6 & 0.3 \\ 0.4 & 0.7 & 0.2 & 0.9 \end{pmatrix}, R_2 = \begin{pmatrix} 0.6 & 0.5 \\ 0.2 & 0.8 \\ 0.9 & 0.3 \\ 0.8 & 0.1 \end{pmatrix},$

solve: $R_1 \circ R_2, R_1 \circ \overline{R_2}.$

19. Given that the matrix of fuzzy relation is:

$$
R = \begin{pmatrix} 1 & 0.2 & 0.5 & 0.1 \\ 0.1 & 0.4 & 0.1 & 0 \\ 0.3 & 0.9 & 0 & 0.4 \end{pmatrix}, \quad S = \begin{pmatrix} 0.4 & 0.9 \\ 0.7 & 1 \\ 0.1 & 0.3 \\ 0.2 & 0.8 \end{pmatrix},
$$

try to solve the matrix of synthetic fuzzy relation $T = R \circ S$.

20. Suppose that $A = \begin{pmatrix} 1 & 0 & 0.3 \\ 0.5 & 0.7 & 0.4 \\ 0 & 1 & 0.5 \end{pmatrix}, B = \begin{pmatrix} 0.2 & 0.4 & 0.8 \\ 0.3 & 0.6 & 0.9 \\ 0.1 & 0.4 & 0.7 \end{pmatrix}$,

try to solve $A \cup B, A \cap B, A \circ B, B \circ A, A^c, A^T$.

21. Suppose that

$$
R_1 = \begin{pmatrix} 0.5 & 0.4 & 0.2 & 0.1 \\ 0.2 & 0.6 & 0.4 & 0.5 \\ 0.1 & 0.9 & 1 & 0.7 \end{pmatrix}, R_2 = \begin{pmatrix} 0.4 & 0.1 & 0.1 & 0.6 \\ 0.5 & 0.9 & 0.7 & 0.5 \\ 0.6 & 0.8 & 0.7 & 0.6 \end{pmatrix}, R_3 = \begin{pmatrix} 0.6 & 0.5 \\ 0.7 & 0.8 \\ 1 & 0.9 \\ 0.2 & 0.3 \end{pmatrix},
$$

$$
R_4 = \begin{pmatrix} 0.7 & 0.3 \\ 0.6 & 0.4 \\ 0.5 & 0.5 \\ 0.4 & 0.6 \end{pmatrix}, R_5 = \begin{pmatrix} 0.2 & 0.3 & 0.7 \\ 0.6 & 0.4 & 0.8 \end{pmatrix}, R_6 = \begin{pmatrix} 0.2 & 0.6 & 0.9 \\ 0.3 & 0.2 & 0.1 \end{pmatrix},
$$

try to solve $(R_1 \cup R_2) \circ (R_3 \circ R_5), (R_1^c \circ (R_3 \cap R_4)) \circ R_6$.

22. Given that the matrix of fuzzy relation is:

$$
R = \begin{pmatrix} 1 & 0.8 & 0 & 0.1 & 0.2 \\ 0.8 & 1 & 0.4 & 0 & 0.9 \\ 0 & 0.4 & 1 & 0 & 0 \\ 0.1 & 0 & 0 & 1 & 0.5 \\ 0.2 & 0.9 & 0 & 0.5 & 1 \end{pmatrix},
$$

try to calculate the second power and the fourth power of R.

Summary

I. The Knowledge Structures of This Chapter

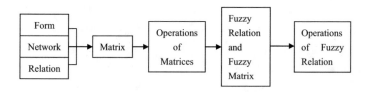

II. Review and Reflection

1. Review that a matrix can be expressed by a form, a network, and a relation.
2. Consider the matrix's addition, subtraction, number-multiplication, and multiplication.
3. A fuzzy relation corresponds to a fuzzy matrix.
4. Solve fuzzy relation's problems by fuzzy matrix operations.

Chapter 3
Fuzzy Control

From Automatic Control to Fuzzy Control

In order to understand fuzzy control, we have to start with automatic controls. There are many examples in automatic controls in our daily lives. We want to use the toilet every day. Its water tank is controlled automatically. We usually use the electric oven, an automatic thermostat controlling the temperature automatically in it. The automatic control technology has gone through a hundred years. If we say that early simple automatic adjustment is controlled by mechanical ways (such as a centrifugal governor), electronic circuitry (such as an oven thermostat). Then the computer has become an indispensable protagonist in the automatic control systems at present. With more widely application of automatic control systems, the controlled objects become more and more complicated, so a type of objects has made designers headache. This kind of controlled objects is very special, which is difficult to establish their mathematical models by commonly used methods, or even a mathematical model is also not accurate enough for it. For this class of controlled objects, the original control technology seems powerless. This problem has troubled engineers for a long time, and finally they found two solutions, one of which is fuzzy control technology.

3.1 Fuzzy Control Full-Automatic Washing Machine

Generally a computer full-automatic washing machine, which is programmed with preset storage programs, is used only by the chosen stored programs. It cannot optimize the selection process all by itself with the change of the number of washing, type, temperature, dirty conditions. Therefore, in this sense, its "full-automatic" function does not have any intelligence while fuzzy logic control full-automatic washing machine has stepped forward to a real intelligent automatic machine. Its goal aims to decide how much water, how much detergent, water intensity and washing time it needs depending on the number requested, texture and

H.-R. Lin et al., *Fuzzy Sets Theory Preliminary*,
https://doi.org/10.1007/978-3-319-70749-5_3

dirtiness. It can dynamically change the parameters to achieve in washing the clothes clean, but it tries not to hurt the clothing with purpose of energy-saving, water-saving and time-saving. It also requires a simple operation by a press of a button to make it work accordingly. Now a full-automatic washing machine with fuzzy logic control technology has been developed to meet these basic conditions for automatic washing machine. The technology continually improves the performance and level of intelligence. In the same conditions, the cleaning efficiency of some fuzzy automatic washing machines has been increased by 20% than that of ordinary washing machines, saving 30% of energy. Its goal is anticipated reaching a smart level of an experienced housewife's intuitive judgment.

3.1.1 Laundry Conditions

To make clothes clean and remove dirt, it depends on the following factors: the clothes materials, water, temperature, detergent performance, mechanical force magnitude, washing time and so on.

(a) **The Materials of Clothes**

Generally the clothing materials can be divided into two categories: natural fiber cotton clothes and man-made chemical fiber clothes. For cotton clothes, the dirt not only attaches upon the fiber surface, but also penetrates into the fibers. So cotton clothes are more difficult to wash than chemical fiber ones.

(b) **Water**

Usually the water can take away dust and water-soluble dirt, so some dirt can be washed off without any detergent. But due to the surface tension, water cannot dissolve or decompose fats and oils dirt. Water hardness can affect washing effect when using soap. But the greatest impact is the temperature of water. Within a certain range, the higher the washing temperature is, the better washing effect it will get. Figure 3.1 is a graph of relation between water temperature and detergency. However, the temperature cannot be too high; otherwise the temperature will attach protein coagulation to the clothes to affect washing effect.

Fig. 3.1 Laundry temperatures (°C)

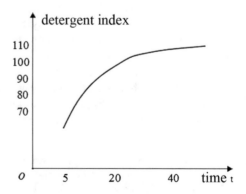

(c) **Detergent**

Detergent ingredients are surfactant-based zeolite, and sodium carbonate etc. Different detergents will be added up with different auxiliary agents, such as enzymes, re-adhesion preventing agents, and optical brighteners etc. In recent years, the detergents include cleanly decontamination protease (proteolytic enzyme), lipase (lipolysis enzyme), cellulase (cellulase enzyme) and other powdered detergent enzymes as an increasingly mainstream.

(d) **The Mechanical Force**

Even if a new detergent containing various high-quality enzymes has been used, the clothes will not be fully cleaned up if they are not moved. In order to make dirt leave clothes surface, you need to bring uniform pressure to bear on the clothes, namely kneading, rubbing, tapping, pressing and other mechanical forces. The larger the mechanical force is, the better the decontamination effect is. But the negative factor is that the mechanical force can also cause damage to the clothes to a certain extent. So it is necessary to set the proper mechanical force (water flow strength and washing time) according to different clothes quantity and quality.

3.1.2 Structure and Sensors of Fuzzy Control Full-Automatic Washing Machine

Figure 3.2 is a cross sectional view of the structure of fuzzy control full-automatic washing machine. It consists of inner barrel, outer barrel, impeller, motor, inlet valve, drain valve and a variety of sensors. To control the washing machine, firstly a variety of sensors have to be used to detect continuously the relevant states or data of the clothes, such as weight, texture, soiling degree, and current water level etc. as a basis for control.

Fig. 3.2 Schematic diagram of the structure cross section

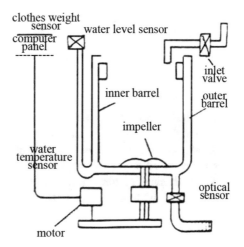

(a) A Laundry Weight Sensor

It is mainly used to detect the laundry weight, in order to determine the water level. Now generally, a dynamic indirect measurement method is used. Because the quantity of laundry in the washing machine will directly affect the motor loads in a washing machine, it is possible to achieve by detecting the motor loads. After putting the laundry in, we start the motor when there is no water or a small amount of water in it, rotating the impeller forward and backward several times. It can detect the amount of clothes according to the voltage-starting time at both ends as the motor stops. The laundry weight sensor works based on this principle. The specific process is to press the "start" button down after putting the clothes in. First water is poured into the washing machine to the water level with a load detector, then pulse circuit drives the motor (0.3 s ON, 0.7 s OFF), it repeats for 32 s totally. The number of pulses transmitted through the optical coupler measures the rotation number with inertia during the motor-off time. When the weight of laundry is large, the number of pulses is small; when the laundry is small, the number of pulses becomes large, whereby the machine can determine the weight of laundry and accordingly determines the water level with one of four speed positions, such as high, medium, low or a little position (Fig. 3.3).

Fig. 3.3 Counting pulses and laundry weight curve

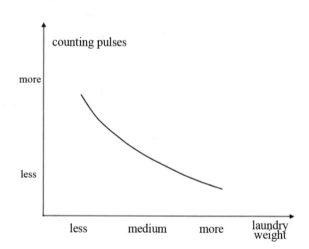

(b) A Cloth Sensor

The cloth detection includes distinguishing cotton from chemical fiber products and differentiating soft cotton cloth from thick cotton cloth. Of course experienced housewife can see it at a glance, but the machine cannot judge it. At the first glance, this is a very difficult problem, but some methods are still shown as follows:

On the basis of earlier detection of the load, drain away a little bit of water. Use the same method to switch the pulse voltage of the drive motor for 32 s with 0.3 s

Fig. 3.4 The distinguishing curves between cotton and chemical fiber clothes

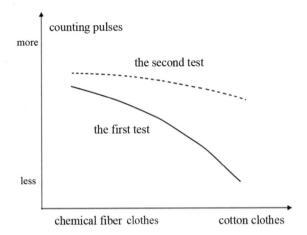

ON and 0.7 s OFF on shift, note down the number N of pulses; if the earlier detected number of pulses with the load is M, then it can determine the general situation of the materials according to M-N values. The more cotton clothes there are, the greater the value of M-N is; and vice versa, the more chemical fiber clothes, the smaller the M-N value. Figure 3.4 shows that the correlative curves between the two tests change with different proportion of cotton and chemical fiber products.

For the same cotton textiles, the washing methods differ as the towels are made of soft cotton textiles and the denims are made of hard thick cotton textiles. How to be distinguished? A water level sensor has to be used to help the test. The specific method is: after the water injection, the pulses drive for 32 s, compare the change before and after the pulses starts. For easy absorbent cotton textiles, when the first pulse signal is issued, almost all cotton textiles have absorbed so much water that no big change would happen when the second pulse signal is issued. So the less change in water level can be judged to be the soft cotton textiles like toweling etc., whereas the denims that are made of dense and thick cotton textiles do not suck enough water when the motor is running for the first time. It would continue to suck water when the motor is running for the second time. So it can be determined to be the hard textures if a large amount of change happens at the water level (Fig. 3.5).

(c) A Water Level Sensor

The water level sensor is an empty pipe contoured with the clothes-washing cylinder that links with the pipe. At the top of the pipe, there is a differential inductrinator separated by a pressure membrane. When water is injected into the cylinder, the air in the pipe is compressed so that the pressure is increased on the pressure membrane, thereby promoting interaction with its iron core movement, causing changes in inductance of the coil.

Using LC oscillator frequency for this inductrinator can be configured to reflect the level of water. For different water level of washing machines, the water level sensor has a corresponding frequency of the pulse signal output (Table 3.1). This

Fig. 3.5 The water level changes curves with soft cloth and hard thick cloth

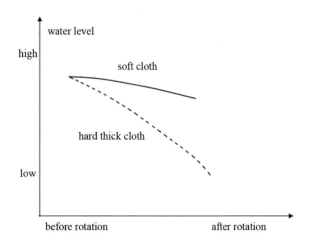

Table 3.1 The input-output relation of a water level sensor

Water level/mm	Frequency/KHz	Water level/mm	Frequency/KHz
450	21.70	200	23.85
400	21.98	150	24.39
350	22.30	100	24.83
300	22.69	50	25.21
250	23.28	0	25.58

sensor can be used to detect more complex fabric softness, but also can be used as a detection device for water level control. In the process of water injection, when the output pulse signal from water level sensor is equal to the selected frequency stored in the singlechip, the singlechip judges the water level has met the requirement. Then the washing machine stops water injection to start the washing program.

(d) A Water Temperature Sensor

The water temperature sensor is installed in the lower part of the cylinder, constituted by a thermistor. The water temperature sensor measures the instant start temperature of the washing machine as room temperature. It tests the water temperature when the water injection has finished. Then the measured temperature is changed into a signal inputted into a singlechip. The water can be heated up to control its temperature in need.

(e) A Turbidity Sensor

The turbidity sensor is used to detect the extent and nature of the dirtiness in clothes. Now a more useful turbidity sensor is the infrared photoelectric sensor, which is arranged in the drain outlet. The light-emitting diode and photodiode are mounted on the opposite of the drain. The light-emitting diode emits light by focusing on the water and the light is received through the water by the photodiode.

The received intensity reflects the transparency of the water, that is, the degree of dirty water. This is an indirect method of measuring the dirt of clothing because the dirt of clothing has positive correlation with the degree of dirty water. In addition, according to the different time required to achieve the same dirtiness, the contaminated clothes can be determined by the nature of "mud pollution" or "oil pollution".

When the washing begins in a state of clean water, the luminous intensity of the light-emitting diode is recorded as an initial set value. At this time, there is high transparency. Along with soil released from the clothes, water gradually becomes muddy, and the transparency decreases. Finally there appears a stable saturation value. The nature and extent of dirt can be known according to the change of transparent rate. The mud dirt is usually separated quickly and enters earlier into the water, so it is shorter for water to change from the transparency into the saturation; while oil dirt is separated relatively slowlier, therefore the change of transparent rate is smaller, and the time reaching a saturation value is longer. Similarly, the solid powder detergent or liquid one can also be known from the change of transparent rate. In general, the saturation value reflects the clothes' dirty level while the time it takes to go from transparency to saturation value reflects the nature of the dirt (Fig. 3.6).

An optical sensor is used to determine the cleanness with water turbidity. When turbidity achieves stability, it means cleanness.

Fig. 3.6 Turbidity change curves. **a** Different degree. **b** Different nature

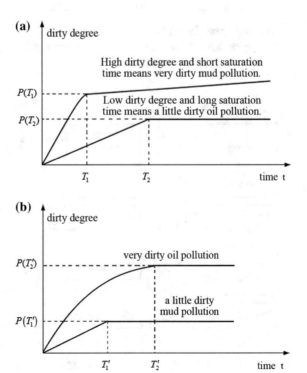

A fuzzy full-automatic washing machine receives the data from its various sensors so as to know the weight, texture, turbidity and type of the laundry to a certain extent. How will it imitate the human brain thinking, reasoning, and make smart decisions to wash clothes with high-efficient energy-saving functions then afterwards? This is the fuzzy control technology to solve the problem.

3.2 A Brief Introduction to Fuzzy Control Theory

Broadly speaking, a fuzzy control is to input the sampling through the sensors and make a fuzzy decision after fuzzy inference to control and manipulate the execution units, in order to complete the automatic control purposes. The fuzzy decision is based on the measurement results, according to some people's experiences summarized as fuzzy rules, after the necessary process of fuzzy mathematics has stored in the computer. The fuzzy rules may have hundreds to thousands of species, and therefore can satisfy choices under a variety of situations. To sum up, fuzzy control experiences have three steps as follows:

1. Input: Input data obtained by the sensors, and fuzz the precise amount up;
2. Operations: Make a fuzzy decision after fuzzy inference by the raw data;
3. Output: Output corresponding control data by comparing with a large number of fuzzy rules previously stored in the database.

A fuzzy washing machine can clean clothes automatically according to the above three steps to simulate the human brain thinking. Figure 3.7 is the control chart of a fuzzy full-automatic washing machine. It uses the testing information of

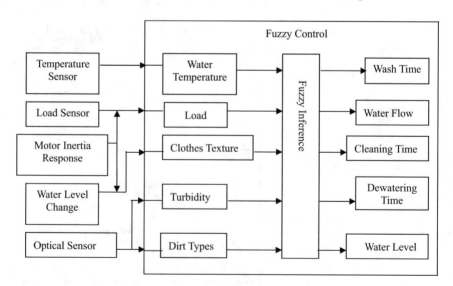

Fig. 3.7 The control chart of fuzzy control washing machine

Fig. 3.8   A fuzzy control full-automatic washing machine

load, texture, water level, water temperature, air temperature and types of detergents etc., makes segmented assessment calculations to fuzz them up, draws a fuzzy inference from them according to the fuzzy rules, then makes a final decision based on the de-activation of the fuzzy rules to determine the most appropriate water level, wash time, cleaning method and dewatering time etc.

Here appear many new terms: fuzziness, fuzzy inference, and defuzzification. A housewife of experiences is easy to know at a glance that these jeans are very dirty and not easy to wash. They have to be washed hard with more detergent and heated water. But how can we make a washing machine "understand and determine" what to do for this arduous task from a lot of data coming from the sensors? And thus "think" how to save both water and electricity, and how to wash them clean—how the machine could imitate the human brain to draw fuzzy inference (Fig. 3.8)?

In general, when an experienced housewife is holding a pile of dirty cotton clothes to clean, she would scrub hard, add some hot water and repeatedly wash several times. That is, her mind has such a fuzzy rule that if you want to wash more clothes and more cotton clothes with much dirt, you should use stronger water flow, higher temperature, and longer washing time; if the circumstances change, such as: less clothes, more chemical fiber clothes, and higher water temperature, then she will have another fuzzy rule: do not wash too hard; use a little detergent; clean up the clothes only; i.e. making the water flow weaker, shortening the wash time, and so on.

For the action principles of foregoing sensors, we can prepare (design) the corresponding control rules on the data of relevant experiences or tests.
For example, the control rules of cloth weight fuzzy inference are:

Rule 1: If the pulse number is very small and the pulse cycle changes largely, then use high water level and preset the long wash time with more detergent.

Rule 2: If the pulse number is small, and the pulse cycle changes moderately, then use medium water level and preset the longer wash time with normal detergent.

...

Follow the same method; we can list each control rule. These control rules are often expressed in tabular form, which is called the control rule table, as shown in Table 3.2.

Table 3.2 The fuzzy inference control rule table for laundry weight

$^v\Delta T$ (pulse cycle change) \ H、T、Q \ P (pulses)		more	general	less	alittle
large	H (waterlevel)	alittle	low	medium	high
large	T (washtime preset)	shorter	normal	longer	long
large	Q (detergent)	normal	normal	more	more
medium	H	alittle	low	medium	high
medium	T	shorter	normal	longer	long
medium	Q	less	less	normal	more
small	H	alittle	low	medium	high
small	T	short	shorter	normal	longer
small	Q	alittle	less	normal	normal

Another example is the fuzzy inference control rules for turbidity:

Rule 1: If the turbidity saturation value is big, the turbidity change rate is slow, and the detergent temperature is low, then the wash time is long on correction.
Rule 2: If the turbidity saturation value is small, the turbidity change rate is fast, and the detergent temperature is high, then the wash time is short on correction.

By the same method, we can design all the fuzzy inference control rules for turbidity, as shown in Table 3.3.

Input variables: turbidity (Tu), the turbidity rate (UT), temperature (θw), the output variables: wash time correction (Δt).

To sum up what an experienced housewife could do, we would have dozens of rules (see Table 3.4).

After the fuzzy control rules have been designed out, some adjustments have to be done to achieve the desired requirements for system output. In the coordination methods of fuzzy control rules, the distribution weighted method is more useful. For example, when you want to make some control rule invalid, you can use the weighted approach to make it failed or reduced intensity.

Table 3.3 All the designed fuzzy inference control rules for turbidity

Tu		Very big			Big			Medium			Small		
θw		High	Medium	Low	High	Medium	Low	High	Medium	Low	High	Medium	Low
UT	Δt												
Fast	Δt t	Bigger	Bigger	Big	Normal	Normal	Normal	Smaller	Smaller	Small	Small	Small	Small
Medium	Δt t	Bigger	Big	Big	Normal	Normal	Bigger	Small	Normal	Normal	Small	Small	Normal
Slow	Δt t	Big	Big	Big	Bigger	Bigger	Big	Normal	Normal	Bigger	Normal	Normal	Normal

Table 3.4 Fuzzy control rules

Texture		More cotton clothes			Half cotton, half chemical fibre clothes			More chemical fibre clothes		
Water temperature		Low	Medium	High	Low	Medium	High	Low	Medium	High
Load	Output									
Big	Water flow	Stronger	Strong	Strong	Strong	Strong	Medium	Medium	Medium	Medium
	Time	Longer	Long	Long	Long	Long	Medium	Long	Medium	Medium
Medium	Water flow	Medium	Medium	Medium	Medium	Medium	Medium	Medium	Weak	Weak
	Time	Long	Medium	Short	Long	Medium	Medium	Medium	Medium	Short
Small	Water flow	Weak	Weak	Weak	Weak	Weak	Weak	Weak	Weak	Weaker
	Time	Medium	Medium	Short	Medium	Short	Short	Medium	Short	Shorter

After repeated experiments and simulations, the engineers have identified 264 washing control methods with different wash time and water flow, by which the following washing controls have been segmented in details:

(1) 9 types of water flow; (2) 16 types of washing time; (3) 6 types of cleaning time; (4) 6 types of dehydrating time; then, in accordance with the method of fuzzy inference, based on fuzzy relation and fuzzy matrix operation, they have been turned into a computer program, and stored in a computer backup.

Now we can divide the program into three steps (Fuzzification of precise amount, Fuzzy inference and Defuzzification of fuzzy amount) to let a fuzzy washing machine decide like human beings to take what ways to wash clothes.

3.3   Fuzzification of Precise Amount

Fuzzification is a transformation process to convert precise amount into fuzzy amount.

The input value that a washing machine collects from a variety of its sensors is precise amount. Every kind of washing states, such as "dirty", "long", "many", "strong" is a fuzzy set. The so-called way to convert accuracy into fuzziness is to turn the input measured value into standardization, that is, the continuous change range of precise amount is mapped into the corresponding discrete domain. Then the input data of the discrete domain is turned into the corresponding language variables terminology, and made into a fuzzy set.

Specifically, if an exact volume corresponds to a subset, this kind of fuzzy set will have an infinite number. For simplicity, accuracy can be discrete, that is, the number can be divided into several intervals, and each interval corresponds to a fuzzy set, so that it would be greatly simplified. If the continuous domain 0–50 is divided into 5 equal parts, we get:

0–10 corresponding element −2
10–20 corresponding element −1
20–30 corresponding element 0,
30–40 corresponding element 1,
40–50 corresponding element 2,

the elements of −2, −1, 0, 1, 2 make a discrete domain.

According to the different control requirements for sensitivity, we usually choose 8 intervals between [−6, 6]. Then the corresponding fuzzy amount can be represented in fuzzy language as follows:

The element in the vicinity of −6 is called "negative-big";
The element in the vicinity of −4 is called "negative-median";
The element in the vicinity of −2 is called "negative-small";
The element slightly less than zero is called "negative-zero";
The element slightly more than zero is called "positive-zero";
The element in the vicinity of +2 is called "positive-small";
The element in the vicinity of +4 is called "positive-median";
The element in the vicinity of +6 is called "positive-big".

Table 3.5 A membership degree corresponds to a fuzzy set of some value

	−6	−5	−4	−3	−2	−1	0	+1	+2	+3	+4	+5	+6
Positive-big	0	0	0	0	0	0	0	0	0.2	0.4	0.7	0.8	1
Positive-median	0	0	0	0	0	0	0	0	0.2	0.7	1	0.7	0.2
Positive-small	0	0	0	0	0	0	0.3	0.8	1	0.7	0.5	0.4	0.2
Positive-zero	0	0	0	0	0	0	1	0.6	0.1	0	0	0	0
Negative-zero	0	0	0	0	0.1	0.6	1	0	0	0	0	0	0
Negative-small	0.2	0.4	0.5	0.7	1	0.8	0.3	0	0	0	0	0	0
Negative-median	0.2	0.7	1	0.7	0.2	0	0	0	0	0	0	0	0
Negative-big	1	0.8	0.7	0.4	0.2	0	0	0	0	0	0	0	0

Their corresponding fuzzy subsets are shown in Table 3.5. The number in the Table represents the corresponding element's membership degree.

The method to fuzz up accuracy is first to input accuracy into [−6, 6] interval values. The values of different fuzzy sets (levels) have different membership degrees. For example, the transformed value 3 corresponds to 0.4 as the membership degree of "positive-big" membership function, 0.7 as the membership degree of "positive-median", 0.7 as the membership degree of "positive-small", and 0 as the membership degree for other fuzzy subsets. It means that any exact value can have access to different degrees of membership functions (that is, suitability). This completes the fuzzification process for accuracy.

In practical calculation, the variable scope may not happen to be in [−6, +6]. The range could be fixed according to their own circumstances. If it is difficult to determine the range, we can change the variable x in the interval $[\alpha, \beta]$ into the variable y in the interval [−6, +6] by the following formula:

$$y = \frac{12}{\beta - \alpha}\left(x - \frac{\alpha + \beta}{2}\right). \tag{3.1}$$

Example 3.1 What does 100 in the [80, 200] correspond to in the [−6, 6]? Solve:

$$y = \frac{12}{200 - 80}\left[100 - \frac{80 + 200}{2}\right]$$
$$= \frac{1}{10} \cdot (-40)$$
$$= -4,$$

i.e., the 100 in the [80, 200] corresponds to −4 in the [−6, 6]. That is, here 100 in the amount of precision is equivalent to the "negative-median" in the fuzzy set. Then we can change it into the corresponding language variables, such as dirty, dirtier, dirtiest, etc. according to the meaning of actual problems.

Example 3.2 If the cloth quality sensor gets the pulse number difference within 80–200 two times, (suppose the "positive-zero" means that nearly half of the laundry is cotton products (slightly more), the "positive-small" means more cotton products, "positive-median" for many cotton products, the "positive-big" for full cotton products; the "negative-zero" for nearly half chemical fiber products (slightly more), the "negative-small" for more chemical fiber products, the "negative-median" for many chemical fiber products, the "negative-big" for full chemical fiber products), so what does the pulse number difference of 100 correspond to for the above eight fuzzy subsets?

Figure by Example 3.1, the pulse difference of 100 corresponds to −4 in the [−6, 6].

From Table 3.5, we find that −4 corresponds to 0.5 as the membership degree for the "negative-small", 1 as the membership degree for the "positive-median", 0.7 as the membership degree for the "negative-big", so the pulse difference of 100 corresponds to the fuzzy subset of "many chemical fiber products".

3.4 Fuzzy Inferences

Fuzzy inferences based on the fuzzy control rules set in advance, can solve the output range of input information by fuzzy relation synthesis algorithm. More generally, the fuzzy relation R can be regarded as a "fuzzy converter" as shown in Fig. 3.9. A is for an input; B for an output. If A and B are given, it wants to solve B.

That is, if we have known the input and the converter, we want to solve the output. This is the problem of fuzzy conversion, which is also called comprehensive evaluation. On the contrary, if B and R are given, it wants to solve A. Then it is the

inverse problem of fuzzy conversion. It can infer the "original reasons" of having made a decision, that is, it is the weighted problem of all factors in the heads. This problem is of great practical significance. It can find out how all kinds of experts, doctors and experienced workers would treat various problems with very valuable but unutterable experiences.

Fig. 3.9 A fuzzy converter

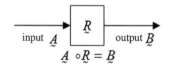

$$A \circ R = B$$

input A · · · output B

As mentioned earlier, the fuzzy control rules are essentially the summary of the people's experiences in the control process. The experiences have been summed up in the following three forms:

① "if A, then B" type, that is,

   IF A THEN B, can be written as $A \to B$.

For example, "If there are many chemical fiber products, then washing time can be a little shorter." is an example of this type.

② "A, then B, or else C" type, that is,

   IF A THEN B, ELSE C, can be written as $A \to B, \bar{A} \to C$.

For example: "If water temperature is OK, stop heating, or else goes on heating" is of this type.

③ "if A and B, then C" type, that is,

   IF A AND B THEN C.

For example: "If there are too many clothes and they are dirty, then the water flow should be strong" is of this type.

The above three forms are the most commonly used fuzzy reasoning in fuzzy control. In these three types of control languages, A, B, C are fuzzy concepts, so they are all relevant fuzzy subsets in domains X, Y and Z.

① For the "if A, then B" type's language, given input, then the output; given input, the output is available to the following formula

$$B_1 = A \circ R \tag{3.2}$$

got, in the formula

$$R = A \times B.$$

This language inference is used for the simple single-input and single-output system, as shown in Fig. 3.10:

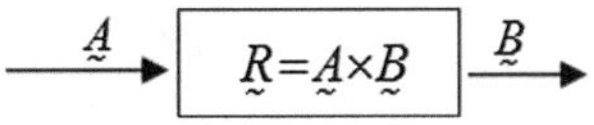

Fig. 3.10 The single-input and single-output system

Example 3.3 Given that when the input $A = 1.0/a_1 + 0.8/a_2 + 0.5/a_3 + 0.2/a_4 + 0/a_5$, the output

$$B = 0.7/b_1 + 1/b_2 + 0.6/b_3 + 0/b_4,$$

Required when the input $A' = 0.4/a_1 + 0.7/a_2 + 1/a_3 + 0.6/a_4 + 0/a_5$, the output $B' = ?$

Solve: First to calculate the fuzzy relation matrix R:

$$R = A \times B = \begin{pmatrix} 1 \\ 0.8 \\ 0.5 \\ 0.2 \\ 0 \end{pmatrix} \circ (0.7 \quad 1 \quad 0.6 \quad 0)$$

$$= \begin{pmatrix} 1.0 \wedge 0.7 & 1.0 \wedge 1.0 & 1.0 \wedge 0.6 & 1 \wedge 0 \\ 0.8 \wedge 0.7 & 0.8 \wedge 1.0 & 0.8 \wedge 0.6 & 0.8 \wedge 0 \\ 0.5 \wedge 0.7 & 0.5 \wedge 1.0 & 0.5 \wedge 0.6 & 0.5 \wedge 0 \\ 0.2 \wedge 0.7 & 0.2 \wedge 1.0 & 0.2 \wedge 0.6 & 0.2 \wedge 0 \\ 0 \wedge 0.7 & 0.0 \wedge 1.0 & 0.0 \wedge 0.6 & 0 \wedge 0 \end{pmatrix}$$

$$= \begin{pmatrix} 0.7 & 1 & 0.6 & 0 \\ 0.7 & 0.8 & 0.6 & 0 \\ 0.5 & 0.5 & 0.5 & 0 \\ 0.2 & 0.2 & 0.2 & 0 \\ 0 & 0 & 0 & 0 \end{pmatrix}.$$

In this way, we can figure B':

$$B' = A' \circ R = (0.4 \quad 0.7 \quad 1 \quad 0.6 \quad 0) \circ \begin{pmatrix} 0.7 & 1 & 0.6 & 0 \\ 0.7 & 0.8 & 0.6 & 0 \\ 0.5 & 0.5 & 0.5 & 0 \\ 0.2 & 0.2 & 0.2 & 0 \\ 0 & 0 & 0 & 0 \end{pmatrix}$$

$= ((0.4 \wedge 0.7) \vee (0.7 \wedge 0.7) \, (0.6 \wedge 0.2) \vee (0 \wedge 0) \, (0.4 \wedge 1)$
$\quad \vee (0.7 \wedge 0.8) \vee (1 \wedge 0.5) \vee (0.6 \wedge 0.2) \vee (0 \wedge 0)$
$\quad (0.4 \wedge 0.6) \vee (0.7 \wedge 0.6) \vee (1 \wedge 0.5) \vee (0.6 \wedge 0.2) \vee (0 \wedge 0) \, (0.4 \wedge 0)$
$\quad \vee (0.7 \wedge 0) \, (1 \wedge 0) \vee (0.6 \wedge 0) \vee (0 \wedge 0))$
$= ((0.4 \vee 0.7 \vee 0.5 \vee 0.2 \vee 0) \, (0.4 \vee 0.7 \vee 0.5 \vee 0.2 \vee 0)$
$\quad (0.4 \vee 0.6 \vee 0.5 \vee 0.2 \vee 0) \, (0 \vee 0 \vee 0 \vee 0 \vee 0))$
$= (0.7 \, 0.7 \, 0.6 \, 0),$

that is

$$B' = 0.7/b_1 + 0.7/b_2 + 0.7/b_3 + 0.7/b_4$$

Example 3.4 Given that the input A is

$$A = \frac{1}{a_1} + \frac{0.8}{a_2} + \frac{0.5}{a_3} + \frac{0.2}{a_4} + \frac{0}{a_5},$$

then the output B is

$$B = \frac{0.8}{b_1} + \frac{1}{b_2} + \frac{0.4}{b_3} + \frac{0}{b_4}.$$

When the input is

$$A_1 = \frac{0.4}{a_1} + \frac{0.8}{a_2} + \frac{1}{a_3} + \frac{0.5}{a_4} + \frac{0}{a_5},$$

the output $B_1 = ?$

Solve: First to solve the fuzzy relation matrix R:

$$R = A \times B = \begin{pmatrix} 1 \\ 0.8 \\ 0.5 \\ 0.2 \\ 0 \end{pmatrix} \circ (0.8 \quad 1 \quad 0.4 \quad 0) = \begin{pmatrix} 0.8 & 1 & 0.4 & 0 \\ 0.8 & 0.8 & 0.4 & 0 \\ 0.5 & 0.5 & 0.4 & 0 \\ 0.2 & 0.2 & 0.2 & 0 \\ 0 & 0 & 0 & 0 \end{pmatrix},$$

then

$$B_1 = A_1 \circ R$$

$$= (0.4 \quad 0.8 \quad 1 \quad 0.5 \quad 0) \circ \begin{pmatrix} 0.8 & 1 & 0.4 & 0 \\ 0.8 & 0.8 & 0.4 & 0 \\ 0.5 & 0.5 & 0.4 & 0 \\ 0.2 & 0.2 & 0.2 & 0 \\ 0 & 0 & 0 & 0 \end{pmatrix}$$

$$= (0.8 \quad 0.8 \quad 0.4 \quad 0),$$

that is

$$B_1 = \frac{0.8}{b_1} + \frac{0.8}{b_2} + \frac{0.4}{b_3} + \frac{0}{b_4}.$$

In fact, the control experiences to a process or an object can be summed up in many items, so that their corresponding inference languages would be expressed in many items, such as

$$IF\ \underset{\sim}{A_1}\ THEN\ \underset{\sim}{B_1},$$
$$IF\ \underset{\sim}{A_2}\ THEN\ \underset{\sim}{B_2},$$
$$\cdots$$
$$IF\ \underset{\sim}{A_n}\ THEN\ \underset{\sim}{B_n},$$

Corresponding to each inference language, it can get relevant fuzzy relation $\underset{\sim}{R_1}, \underset{\sim}{R_2}, \ldots, \underset{\sim}{R_n}$, so the $\underset{\sim}{R}$ corresponded to total control rules of the system can be obtained by Union to "take the bigger" as follows

$$\underset{\sim}{R} = \underset{\sim}{R_1} \cup \underset{\sim}{R_2} \cup \cdots \cup \underset{\sim}{R_n} = \cup_{i=1}^{n} \underset{\sim}{R_1}. \tag{3.3}$$

② For the "if $\underset{\sim}{A}$, then $\underset{\sim}{B}$, or else $\underset{\sim}{C}$" (If $\underset{\sim}{A}$ Then $\underset{\sim}{B}$ Else $\underset{\sim}{C}$) type's fuzzy rules, the following inference method may be adopted.

Given that the input is $\underset{\sim}{A}$, the output is $\underset{\sim}{B}$, or else the output is $\underset{\sim}{C}$.

Now given that the input is $\underset{\sim}{A}$, the output is $\underset{\sim}{B}$, or $\underset{\sim}{C}$, which can be obtained from the following equation:

$$\underset{\sim}{B'} = \underset{\sim}{A'} \circ R_1,$$
$$\underset{\sim}{C'} = \underset{\sim}{\bar{A}'} \circ R_2,$$

of which, the fuzzy relation $R_1 = \underset{\sim}{A} \times \underset{\sim}{B},\quad R_2 = (\underset{\sim}{\bar{A}} \times \underset{\sim}{C}).$

Example 3.5 Suppose domain X = Y={1, 2, 3, 4, 5}, $\underset{\sim}{A} \in X$, $\underset{\sim}{B} \in Y$, and

$$\underset{\sim}{A} = [black] = \frac{1}{1} + \frac{0.5}{2} + \frac{0.1}{3},$$

$$\underset{\sim}{B} = [white] = \frac{0.3}{3} + \frac{0.8}{4} + \frac{1}{5}.$$

Solve the fuzzy relation of "if x is black, then y is white, or y is not very white".

Solve: As "very" is a fuzzy operator, so $\underset{\sim}{C}$ = [not very white] = 1 − [very white] = $1 - \underset{\sim}{B}^2$, that is,

$$C = \frac{1}{1} + \frac{1}{2} + \frac{0.91}{3} + \frac{0.36}{4} + \frac{0}{5},$$

$$R_1 = A \times B = \begin{pmatrix} 1 \\ 0.5 \\ 0.1 \\ 0 \\ 0 \end{pmatrix} \circ (0 \quad 0 \quad 0.3 \quad 0.8 \quad 1) = \begin{pmatrix} 0 & 0 & 0.3 & 0.8 & 1 \\ 0 & 0 & 0.3 & 0.5 & 0.5 \\ 0 & 0 & 0.1 & 0.1 & 0.1 \\ 0 & 0 & 0 & 0 & 0 \\ 0 & 0 & 0 & 0 & 0 \end{pmatrix},$$

$$\bar{A} = 1 - A = \frac{0}{1} + \frac{0.5}{2} + \frac{0.9}{3} + \frac{1}{4} + \frac{1}{5},$$

$$R_2 = \bar{A} \times C = \begin{pmatrix} 0 \\ 0.5 \\ 0.9 \\ 1 \\ 1 \end{pmatrix} \circ (1 \quad 1 \quad 0.91 \quad 0.36 \quad 0) = \begin{pmatrix} 0 & 0 & 0 & 0 & 0 \\ 0.5 & 0.5 & 0.5 & 0.36 & 0 \\ 0.9 & 0.9 & 0.9 & 0.36 & 0 \\ 1 & 1 & 0.91 & 0.36 & 0 \\ 1 & 1 & 0.91 & 0.36 & 0 \end{pmatrix}.$$

So from $R = R_1 \cup R_2$, we get

$$R = \begin{pmatrix} 0 & 0 & 0.3 & 0.8 & 1 \\ 0.5 & 0.5 & 0.5 & 0.5 & 0.5 \\ 0.9 & 0.9 & 0.9 & 0.36 & 0.1 \\ 1 & 1 & 0.91 & 0.36 & 0 \\ 1 & 1 & 0.91 & 0.36 & 0 \end{pmatrix}.$$

R is the fuzzy relation of "if x is black, then y is white, or y is not very white".

③ The corresponding fuzzy controller of "if A and B, then C" type language is a control model of double-input and single-output type, whose structure is shown in Fig. 3.11.

A general fuzzy controller would adopt this double-input and single-output model. That is, it not only needs to use the error of controlled amount as control information in the control process, but also it is required to use the error change rate as control information. In this way, a better control effect could be achieved.

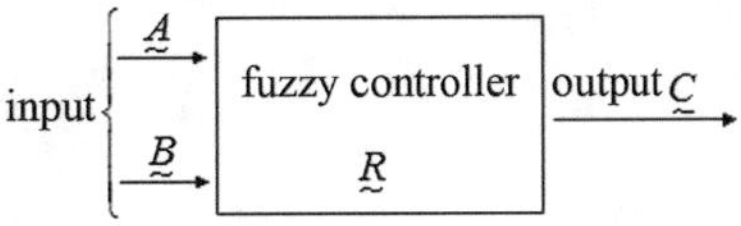

Fig. 3.11 The fuzzy control model of double-input and single-output model

Example 3.6 Given that when the input $E = \dfrac{0.5}{e_1} + \dfrac{1}{e_2}$, $C = \dfrac{0.1}{c_1} + \dfrac{1}{c_2} + \dfrac{0.6}{c_3}$, its

output $U = \dfrac{0.4}{u_1} + \dfrac{1}{u_1}$, solve the fuzzy relation R corresponding to this rule.

Solve: This double-input and single-output (A and $B \rightarrow C$) model is of multi-relation, its fuzzy relation matrix can be obtained by Madani inference.

First solve $D = E \times C$,

$$
\begin{aligned}
D &= \begin{pmatrix} 0.5 \\ 1 \end{pmatrix} \circ \begin{pmatrix} 0.1 & 1 & 0.6 \end{pmatrix} \\
&= \begin{pmatrix} 0.5 \wedge 0.1 & 0.5 \wedge 1 & 0.5 \wedge 0.6 \\ 1 \wedge 0.1 & 1 \wedge 1 & 1 \wedge 0.6 \end{pmatrix} \\
&= \begin{pmatrix} 0.1 & 0.5 & 0.5 \\ 0.1 & 1 & 0.6 \end{pmatrix}.
\end{aligned}
$$

Then rewrite D into the following form:

$$
D^T = \begin{pmatrix} 0.1 \\ 0.5 \\ 0.5 \\ 0.1 \\ 1 \\ 0.6 \end{pmatrix}.
$$

D^T is a transpose of D, where the elements in the first row is written in the column order, then the elements in the second row is written in the column order following the first row; elements in multiple rows are followed in turn and so on. Hence

$$
R = D^T \times U = \begin{pmatrix} 0.1 \\ 0.5 \\ 0.5 \\ 0.1 \\ 1 \\ 0.6 \end{pmatrix} \circ \begin{pmatrix} 0.4 & 1 \end{pmatrix} = \begin{pmatrix} 0.4 \wedge 0.1 & 1 \wedge 0.1 \\ 0.4 \wedge 0.5 & 1 \wedge 0.5 \\ 0.4 \wedge 0.5 & 1 \wedge 0.5 \\ 0.4 \wedge 0.1 & 1 \wedge 0.1 \\ 0.4 \wedge 1 & 1 \wedge 1 \\ 0.4 \wedge 0.6 & 1 \wedge 0.6 \end{pmatrix} = \begin{pmatrix} 0.1 & 0.1 \\ 0.4 & 0.5 \\ 0.4 & 0.5 \\ 0.1 & 0.1 \\ 0.4 & 1 \\ 0.4 & 0.6 \end{pmatrix}.
$$

Now we have the fuzzy relation R, if given the inputs E' and C', the output U' can be calculated:

First calculate

$$D' = E' \times C',$$

then solve

$$U' = D'^{\mathrm{T}} \circ R.$$

Example 3.7 In the above problem, if

$$E' = \frac{1}{e_1} + \frac{0.5}{e_2}, C' = \frac{0.1}{c_1} + \frac{0.5}{c_2} + \frac{1}{c_3},$$

solve the output U'.

Solve: Because

$$D' = E' \times C' = \begin{pmatrix} 1 \\ 0.5 \end{pmatrix} \circ (0.1 \quad 0.5 \quad 1)$$

$$= \begin{pmatrix} 0.1 & 0.5 & 1 \\ 0.1 & 0.5 & 0.5 \end{pmatrix},$$

$$D'^{\mathrm{T}} = \begin{pmatrix} 0.1 \\ 0.5 \\ 1 \\ 0.1 \\ 0.5 \\ 0.5 \end{pmatrix},$$

then

$$U' = D'^{\mathrm{T}} \circ R = (0.1 \quad 0.5 \quad 1 \quad 0.1 \quad 0.5 \quad 0.5) \circ \begin{pmatrix} 0.1 & 0.1 \\ 0.4 & 0.5 \\ 0.4 & 0.5 \\ 0.1 & 0.1 \\ 0.4 & 1 \\ 0.4 & 0.6 \end{pmatrix}$$

$$= (0.4 \quad 0.5).$$

So the control amount of fuzzy output

$$U' = \frac{0.4}{u_1} + \frac{0.5}{u_2}.$$

Now, if the washing machine receives a set of data from the sensor: 1, 0.8, 0.5, 0.2, 0, it means standard water temperature, many clothes, half cotton and half chemical fiber products, not too dirty, no water respectively. Then we assume that according to some activated fuzzy control rules by fuzzy inference operation, it would get a group of new data 0.7, 1, 0.6, 0.4, 0.6, which means longer washing time, standard cleaning time, medium slightly stronger flow, less dehydration time, medium water level. This is the fuzzy decision that the washing machine has made. In this way, it can both wash clean, and save water and electricity. How to solve the fuzzy control rules?

This is a typical "if A, then B" type. According to $R = A \times B$, we can get

$$R = A \times B = \begin{pmatrix} 1 \\ 0.8 \\ 0.5 \\ 0.2 \\ 0 \end{pmatrix} \circ \begin{pmatrix} 0.7 & 1 & 0.6 & 0.4 & 0.6 \end{pmatrix}$$

$$= \begin{pmatrix} 1.0 \wedge 0.7 & 1.0 \wedge 1.0 & 1.0 \wedge 0.6 & 1 \wedge 0.4 & 1 \wedge 0.6 \\ 0.8 \wedge 0.7 & 0.8 \wedge 1.0 & 0.8 \wedge 0.6 & 0.8 \wedge 0.4 & 0.8 \wedge 0.6 \\ 0.5 \wedge 0.7 & 0.5 \wedge 1.0 & 0.5 \wedge 0.6 & 0.5 \wedge 0.4 & 0.5 \wedge 0.6 \\ 0.2 \wedge 0.7 & 0.2 \wedge 1.0 & 0.2 \wedge 0.6 & 0.2 \wedge 0.4 & 0.2 \wedge 0.6 \\ 0 \wedge 0.7 & 0.0 \wedge 1.0 & 0.0 \wedge 0.6 & 0 \wedge 0.4 & 0 \wedge 0.6 \end{pmatrix}$$

$$= \begin{pmatrix} 0.7 & 1 & 0.6 & 0.4 & 0.6 \\ 0.7 & 0.8 & 0.6 & 0.4 & 0.6 \\ 0.5 & 0.5 & 0.5 & 0.4 & 0.5 \\ 0.2 & 0.2 & 0.2 & 0.2 & 0.2 \\ 0 & 0 & 0 & 0 & 0 \end{pmatrix}.$$

By experiment comparison, a washing machine with fuzzy control at 5 °C can improve 20% of the cleanness than other average washing machines and can avoid excessive washing of clothes against damages, or fail to fully wash the dirt away, which could save time and energy.

3.5 Defuzzification of Fuzzy Amount

The fuzzy control decision through fuzzy inference cannot directly manipulate the implementation parts because the data it represents is still fuzzy amount. How long is the longer wash time? How much is the medium slightly stronger flow? Any components can only be controlled precisely, so there is a problem of how to turn fuzziness into precise amount. Here are two commonly used methods of defuzzification.

The first method: choosing the maximum membership degree principle, i.e. the element u^* that has the largest membership degree in the fuzzy subsets is selected as the implementation amount. If the corresponding fuzzy set of fuzzy decision is $\underset{\sim}{C}$, then the accuracy that the decision has determined should meet:

$$\mu_c(u^*) \geq u_c(u), \ u \in U. \tag{3.4}$$

If there are a few maximum points: u*, then take their mean as the implementation amount.

Example 3.8 If $\underset{\sim}{C} = \dfrac{0.2}{2} + \dfrac{0.7}{3} + \dfrac{1}{4} + \dfrac{0.7}{5} + \dfrac{0.2}{6}$, then it should take $\mu = 4.$ as the implementation amount according to the maximum membership degree principle.

If there are two of the biggest points: u*, such that $\underset{\sim}{C} = \dfrac{0.1}{-4} + \dfrac{0.4}{-3} + \dfrac{0.8}{-2} + \dfrac{1}{-1} + \dfrac{1}{0} + \dfrac{0.4}{1}$, you should take

$$u^* = \frac{0-1}{2} = -0.5.$$

This judgment method is simple and easy, in real-time on the computer, but it has summed up too little information, and lost a lot of good information in the decision process. This is because the process did not take into account the rest of the judgment of those small membership degrees, and there is no difference in width and distribution of the membership functions.

The second method: the weighted average method: the implementation amount is decided by the following formula

$$u^* = \frac{\sum\limits_{i=1}^{n} \mu(u_i)u_i}{\sum\limits_{i=1}^{n} \mu(u_i)}. \tag{3.5}$$

Example 3.9 If $\underset{\sim}{C} = \dfrac{0.2}{2} + \dfrac{0.7}{3} + \dfrac{1}{4} + \dfrac{0.7}{5} + \dfrac{0.2}{6}$, then

$$u^* = \frac{2 \times 0.2 + 3 \times 0.7 + 4 \times 1 + 5 \times 0.7 + 6 \times 0.2}{0.2 + 0.7 + 1 + 0.7 + 0.2} = 4.$$

It is similar to look for mathematical expectation in probability theory that the weighted average judgment would be used to determine the implementation amount of output information. The decision of weighted coefficient directly affects the response characteristics of the system. Adjusting the weighted coefficient can improve the system's dynamic characteristics, which is very important to the design of fuzzy controller.

In short, the fuzzy decision is directly based on the current input conditions to activate some fuzzy rules set in advance and stored in computer's memory to calculate the membership degree, then using fuzzy algorithm to calculate precise control output.

In this way, the washing machine can use fuzzy logic control technology to realize intelligent controls, so it is with other complex industrial processes. We could start with qualitative analysis of the control of an industrial process.

The first step, the people with eyes, ears and other senses get the fuzzy information of system output amount and its change rate. Of course, this amount objectively is not fuzzy but precise figures, which is reflected in the human mind to become fuzzy amount (such as high temperature, low pressure etc.). Objective accuracy reaches the brain by human senses, and the process is actually the fuzzification process of precise amount.

The second step, based on the fuzzy information and compared with the experiences already stored in mind, people would analyze and judge what control measures they should be taken to the system, that is, determine what input amounts should be adjusted.

The experiences of the operators can be summed up into a number of rules, which are stored into the computer after some mathematical treatment. These rules are called fuzzy control rules. And the computer would make a fuzzy decision based on the input fuzzy information and the control and inference rules.

The third step, when people implement a specific control action based on a fuzzy decision, it needs an exact amount. Though it is a fuzzy decision in mind, such as "turning down the valves", but in fact, people would turn down the valves by the precise specific number. So here again there is a process of a fuzzy value into a precise value.

When people make some operations, they would have finished them by their own experiences and unconsciously according to the above "accurate—fuzzy—accurate" conversion process. But if you want to use some automated device to replace human beings' control, this transformation between fuzzy and accurate amount is essential because the device itself cannot think. So, the fuzzy logic control technology make machine simulate human experiences, undergo the same thinking, and make appropriate decisions.

Fuzzy logic has widely been used and has attracted rave reviews, but fuzzy logic technology is not a universal solution, it also has shortcomings:

1. The fuzzy control still needs further researches in fuzzy modeling, fuzzy rules and fuzzy inference algorithms in the application of non-linear complex system.
2. Due to the interaction of complex fuzzy rules, it makes the gained synthetic inference algorithms possess nonlinear performance, so that fuzzy control would not have considerable ideal effect.
3. The theory of stability for fuzzy control systems needs to be further discussed.
4. The self-study fuzzy control strategies, the intelligent system structure and its implementation still need to continue their efforts.

These shortcomings need the new generation of students to explore and perfect with innovative and pioneering spirit, and need your involvement, advice and contribution.

In 1989, Japan proposed a research project of "Future Direction of Fuzzy Systems" (Fig. 3.12).

There was an article: "It isn't far away for the computers to surpass the human brain." in Liberation Daily in 1999. There are some data and schedules that would excite heart and cause people to think deeply. The excerpts are shown below.

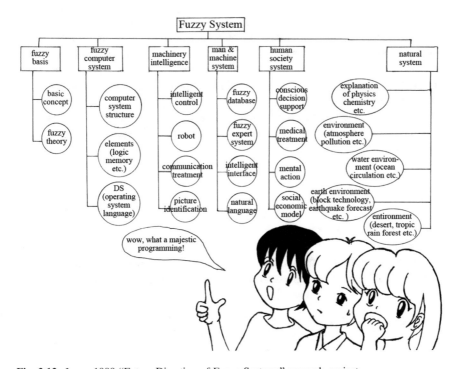

Fig. 3.12 Japan 1989 "Future Direction of Fuzzy Systems" research project

Scientists predict that by 2019, when computer chips speed up to 20 trillion times per second, the machine can sense human emotions; in 2029, with electronic chip transplantation, the blind can see the light, and the deaf can hear the voice; in 2099, robots will be on an equal footing with humans. By then, the CAM will have 1 trillion (10^{12}) neurons, beyond the general level of intelligence of the human brain. This is the stage that people are most worried about. Will humans have the ability to harness this artificial intelligence because it is smarter than the average person?

In spring 1999, the famous physicist Stephen Hawking of Cambridge University, known as the contemporary Einstein, gave a speech at The Welcome Party for New 2000 in White House, Washington: ... Life on Earth seemed to originate from 4 billion years ago, after about 500 million years; there appeared the basis for all life on Earth—a highly complex molecule of DNA. Biological evolution has been very slow. In the last 10,000 years, no change in human DNA, however, over the next 100 years, we will probably be able to redesign the human DNA completely. We are at the beginning of a new era. In this new era, we will be able to increase the complexity of human DNA without having to wait for the slow process of evolution. Whether we like it or not, it may improve human genetic gene engineering.

To some extent, if humanity is to cope with the growing complexity of the world around, and meet new challenges like space travel, they need to improve their mental and physical qualities. If the biological system stays ahead of the electronic system, it also needs to increase its complexity. At present, the machine has a speed advantage, but they would show no signs of intelligence. The speed and complexity of the computer doubles every 128 months, it would develop to the complexity of the human brain. Some people say that the computer no matter what it might be like has never demonstrated real smart. But it seems to me that if very complex chemical molecules acted in the human body make humans intelligent, then equally complicated electronic circuits would cause the computer to act in an intelligent way. If the computer has intelligence, it could design higher complexity and more intelligent computers.

He concluded his speech, quoted a famous saying: "I have seen the future, and it is running."

Exercise 3

1. If $X = \left\{ \underset{\sim}{R} = \underset{\sim}{A} \times \underset{\sim}{B} \right\} = \{1.4, 1.5, 1.6, 1.7, 1.8\}$, which represents a domain of the women height in a certain area, with m as the unit; $Y = \{Y = y_1, y_2, y_3, y_4, y_5\} = \{40, 50, 60, 70, 80\}$ denotes a domain of the woman's weight in that area, with kg as the unit. Also suppose that the set of the concept of "height" is represented as follows to the women in the area,

$$[\text{Height}] = \frac{0.2}{1.4} + \frac{0.5}{1.5} + \frac{0.8}{1.6} + \frac{1}{1.7} + \frac{1}{1.8},$$

the set of the concept of "weight" is represented as follows,

$$[\text{Weight}] = \frac{0.2}{40} + \frac{0.6}{50} + \frac{0.9}{60} + \frac{1}{70} + \frac{1}{80},$$

try to solve the fuzzy relation R to y of "very high" if x is "very high.
2. Suppose that x represents temperature, and y voltage. After they are discrete, the domains of the temperature and the voltage are X = Y= {1, 2, 3, 4, 5}. Given that the fuzzy subsets are on X and Y:

$$\underset{\sim}{A} = [\text{low}] = \frac{1}{1} + \frac{0.7}{2} + \frac{0.3}{3},$$

$$\underset{\sim}{B} = [\text{high}] = \frac{0.4}{3} + \frac{0.7}{4} + \frac{1}{5},$$

$$\underset{\sim}{A'} = [\text{lower}] = \text{low}^{0.75} = \frac{1}{1} + \frac{0.76}{2} + \frac{0.4}{3} + \frac{0.2}{4}.$$

Solve: What is the control voltage when the temperature is lower now, following the inference rule that "If the temperature is low, then the control voltage is high"?
3. An electrothermal drying furnace is controlled artificially for continuous regulation of outer voltage, in order to overcome the interference to reach constant temperature drying purposes. The operators' experiences are that "If its temperature is low, then the outer voltage is high, or the voltage is not very high." Now, try to determine how the outer voltage should be adjusted if the temperature is very low?

Suppose that x represents temperature and y indicates voltage. The above problem can be described as "If x is low, y is high, or y is not very high." How is y if x is very low?

Suppose that domain X = Y= {1, 2, 3, 4, 5},

$$\underset{\sim}{A} = [\text{low}] = \frac{1}{1} + \frac{0.8}{2} + \frac{0.6}{3} + \frac{0.4}{4} + \frac{0.2}{5},$$

$$\underset{\sim}{B} = [\text{high}] = \frac{0.2}{1} + \frac{0.4}{2} + \frac{0.6}{3} + \frac{0.8}{4} + \frac{1}{5},$$

$$\underset{\sim}{C} = [\text{not very high}] = \frac{0.96}{1} + \frac{0.84}{2} + \frac{0.64}{3} + \frac{0.36}{4} + \frac{0}{5},$$

$$\underset{\sim}{A_1} = [\text{very low}] = \underset{\sim}{A}^2 = \frac{1}{1} + \frac{0.64}{2} + \frac{0.36}{3} + \frac{0.16}{4} + \frac{0.04}{5}.$$

4. Suppose that $X = Y = \{1, 2, 3\}$, and

$$A = [\text{small}] = (1, \quad 0.4, \quad .0), \quad B = [\text{big}] = (0, \quad 0.4, \quad 1).$$

If x is small, y is big, or y is not very big. Try seeking a fuzzy relation R.

5. Given that the input $A = \frac{1}{x_1} + \frac{0.5}{x_2}$ and $B = \frac{0.1}{y_1} + \frac{0.5}{y_2} + \frac{1}{y_3}$, then the output $C = \frac{0.2}{z_1} + \frac{1}{z_2}$, according to this rule, determine the output C' when the input

$$A' = \frac{0.8}{x_1} + \frac{0.1}{x_2} \text{ and } B' = \frac{0.5}{y_1} + \frac{0.2}{y_2} + \frac{0}{y_3}.$$

6. Suppose that the domain: $X = \{a_1, a_2, a_3\}, Y = \{b_1, b_2, b_3\}, Z = \{c_1, c_2\}$,

$$A = \frac{0.5}{a_1} + \frac{1}{a_2} + \frac{0.1}{a_3}, A \in X,$$

Given that the fuzzy sets: $B = \frac{0.1}{b_1} + \frac{1}{b_2} + \frac{0.6}{b_3}, B \in Y,$

$$C = \frac{0.4}{c_1} + \frac{1}{c_2}, C \in Z.$$

Try to solve the fuzzy relation R determined by the fuzzy conditional statement "if A and B, then C" and calculating the output fuzzy set C_1 determined by the given input fuzzy sets

$$A_1 = \frac{1}{a} + \frac{0.5}{a_2} + \frac{0.1}{a_3} \text{ and } B_1 = \frac{0.1}{b_1} + \frac{0.5}{b_2} + \frac{1}{b_3}.$$

7. Suppose that the control is a two-input and single-output module, the controlling rules are constituted by the following conditional statement x y z.

If $x = A_1$ and $y = B_1$ then $z = C_1$,
If $x = A_2$ and $y = B_2$ then $z = C_2$,
here the domain $X = \{x_1, x_2\}, Y = \{y_1, y_2, y_3\}, Z = \{z_1, z_2\}$,

$$A_1 = \frac{0.5}{x_1} + \frac{1}{x_2}, \qquad A_2 = \frac{1}{x_1} + \frac{0.5}{x_2},$$

$$B_1 = \frac{0.1}{y_1} + \frac{1}{y_2} + \frac{0.6}{y_3}, \qquad B_1 = \frac{0.6}{y_1} + \frac{1}{y_2} + \frac{0.1}{y_3},$$

$$C_1 = \frac{0.4}{z_1} + \frac{1}{z_2}, \qquad C_2 = \frac{1}{z_1} + \frac{0.4}{z_2}.$$

Try to solve the fuzzy relation R, and solve the output C^* when the inputs

$$A^* = \frac{1}{x_1} + \frac{0.5}{x_2}, \quad B^* = \frac{0.1}{y_1} + \frac{0.5}{y_2} + \frac{1}{y_3}.$$

8. Assuming that the washing machine in the case of the low temperature, needs to wash a lot of dirty clothes made of chemical fibers. Before the water in-pouring, it activates a control rule

$$R = \begin{pmatrix} 0.7 & 1 & 0.6 & 0.4 & 0.6 \\ 0.7 & 0.8 & 0.6 & 0.4 & 0.6 \\ 0.5 & 0.5 & 0.5 & 0.4 & 0.5 \\ 0.2 & 0.2 & 0.2 & 0.2 & 0.2 \\ 0 & 0 & 0 & 0 & 0 \end{pmatrix}.$$

Question: How will it clean the clothes? (How long is the wash time? How strong is the water currents? How long is the dewatering time? How high is the water level?) (assuming that the fuzzy data of the washing clothes are (0.4, 0.7, 1, 0.6, 0), representing water temperature, clothes quantity, materials, dirty level, and water level).

Reading and Reflecting 1

DNA Computers Will "Compete" with Conventional Computers

In November 2001, Israeli scientists developed a computer of minimal deoxyribonucleic acid (DNA). How big is it? One trillion such computers can fill up a test tube. One trillion such computers can calculate at one billion times per second, with 99.8% of accuracy.

This computer is not using numbers and formulas to solve problems. Its input/output devices and software are made up of DNA molecules that are stored in the active organisms and are handling encoding information. Scientists believe that this DNA computer will become a competitor against other more conventional computers henceforth as miniaturization has been stretched to the limit, and DNA computers can calculate much faster than conventional computers.

Professor Ehud Shapiro at Israeli Weizmann Institute, said: "we have created a nano-computer with biological molecules. This computer is very small and cannot be run individually. If one trillion computers could calculate at the same time, they would be able to make one billion operations.

Scientists found that DNA has a characteristic, capable of carrying a large number of genetic materials that the organisms' various cells have. Mathematicians, biologists, chemists and computer specialists get inspired, are working together to develop future liquid DNA computers. The DNA computers work, based on the instantaneous chemical reactions, through interactions with enzymes, and make the reactive process molecular-coding, then let the problems be answered in new DNA coding forms.

First, compared with ordinary computers, the advantage of DNA computers is small in size, but the amount of information stored is more than that of all the computers in the modern world. Its room to store information is only about one trillionth smaller than that of ordinary computers. The information can be stored in a chain of hundreds of thousands of DNA. One liter of DNA computers can finish the operations that all computers have done so far within a few days. Secondly, it can reduce energy consumption at most. The energy consumption of a DNA computer is only one billionth of ordinary computers.

It is because each chain is a micro-processor that a DNA computer is powerful. Scientists can put 1 billion billion chains in 1000 g of water, and each chain can do its own things. They can do their calculations all by themselves. This means that DNA computers can "try" a tremendous amount of solutions at the same time. Contrasted with this, electronic computers must calculate each solution all the way, until the trial of the next program.

Therefore, electronic computers are totally different from DNA computers. An electronic Computer is capable of operating many times within an hour, but it only can make one operation each time while a DNA computer needs about an hour for one operation, but it can make 1 billion operations each time.

The functions of a human brain are somewhere in between: around 10 trillion operations per hour. A DNA computer translates a binary number into genetic code fragments. Each fragment is one of the famous double helix chains. Scientists hope to decompose all possible models of DNA and put it in a test tube. Then they will make complementary digital chains. The complementary digital chain won't solve a certain mode, but they will be extracted from a solution.

The DNA biological computer is the whim brought forward by Dr. Adelaman at University of Southern California in the United States of America in 1994. It completes the operation by controlling the biochemical reaction between DNA molecules. But for the current DNA computer technology, DNA must be dissolved in the liquid of a test tube. This kind of computer is made up of a bunch of tubes filled with organic liquids, and its structure has to be improved yet.

This is the first world independent programmable computer whose input and output equipment and software and hardware are made up all by biological molecules. Although the computer is too simple and it is no direct use currently in commerce, it has high potential scientific values and lays the foundation for researching the computer that can be put and run in the human body, and in the human cell. The latter can act as a monitoring device to find potential pathogenic changes, and also can synthesize medicines in the human body to treat various difficult diseases.

(Jiefang Daily, March 15, 2002)

Editor's Note: The fundamental differences between biological and non-biological organisms are that biological organisms can reproduce, but non-biological organisms cannot. Can a robot be alive? What does it mean as it can replicate itself? Please read the following article published in 2005:

Reading and Reflecting 2

The Robot Can Replicate Itself

The scientists at Cornell University in the USA have recently verified that a robot can replicate itself and heal itself. It is said that the NASA is interested in this technology and is looking forward to self-replicating robots in space to show their talents in the future.

The "self-replicating" is defined as the fact that a machine can produce another same machine in accordance with its own situation. And its replica can continue to reproduce another machine, and can repeat the cycles.

The self-replicating robot that scientists made at Cornell University in the USA looks more like a nursery toy, rather than a Star Wars robot. It consists of 4 intelligent modules. It currently cannot walk or speak. But the inventor of the robot says that he has proved that the machine can achieve self-replication by programming. This is a concept breakthrough, i.e., a robot could survive independently when leaving the human environments.

The intelligent modules that have composed a robot are 10 cm^3 in size, and the modules can be freely rotated 120°. The outer surface of a module is fixed with a magnet, so the modules can be combined or divided by the magnetic strength. The robot in this design can be freely assembled into various shapes, such as towers, rectangular shapes, squares, and so on. Each module has a tiny computer chip, which contains specific instructions when assembled.

The "self-replicating" robot has been the "prop" in the science fiction, but in real life, it seems so out of reach both from technology as well as from software. So far, only 2 machines do make themselves copied.

Mr Hod Lipson who led the research said that though the robot still had many limitations, it had proved at least that it would be possible for a machine to do self-replication, not only the organism could do.

Mr Hod Lipson said: "The self-replicating, self-repairing robot can assemble a new machine even in the environments where human beings can't live. It is the most exciting prospect to do it in space, such as Mars-detecting."

According to Lipson's speech, the study has been widely supported. One of the financial supporters is the NASA in the USA. He said: "People are growing interest in this, they want the robot that could diagnose the problem, and have self-healing high performance. For example, if a Mars-detecting robot got out of work, you might want to fix it on the spot. People don't want to give up the whole detecting task because the robot has a little problem."

Moreover, the technology that the robot replicates and repairs itself can be applied not only in space, but in war zones or in damaged nuclear reactors, where man is not likely to survive, but the robot can.

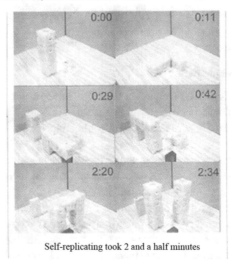

Self-replicating took 2 and a half minutes

Self-replicating took 2 and a half minutes

The robot's self-replication steps were: first, the robot bent and placed a module on the table as the "head" of his copy. Then it turned to another side, and drew a new module on the "head" of his replica with electromagnet attraction. The robot turned these modules according to the requirements, duplicated in accordance with the steps above. In the end, a new robot similar to the original robot in shape was combined successfully. The whole process took 2 min and 34 s. (see the picture below)   (Morning News, May 13, 2005)

Reading and Reflecting 3

"Once I Think I Am Making the God, I Cannot Help Exciting"

An Interview Excerpt of Professor Hugo de Garis, "Father of Artificial Brain"

Professor Hugo de Garis, a world's top scientist, known as "Father of Artificial Brain", who created "a research field of evolutionary hardware and quantum computation, called father of the EH field, father of evolutionary hardware and evolutionary engineering".

He was born in 1947 in Sydney, Australia, and had been studying very prominently at high school. His name has been engraved on the glorious columns at school. At the age of 23, he graduated from University of Melbourne, and obtained two honored degrees: Bachelor of "Applied Mathematics" and Bachelor of "Theoretical Physics", indicating that he is excellent in both disciplines. Then he went to study and work at Cambridge University in the UK. After a few years of working in the computer industry and feeling tired and frustrated, he recognized his life pursuit. He wanted rational life and academic atmosphere. He chose to return to University of Brussels where he had studied, to study for a doctorate in artificial intelligence and artificial life, becoming a researcher. He has worked and studied in seven countries and is fluent in four languages. In 1992, Hugo received Postdoctoral Fellowship in artificial intelligence, leaving from Europe to Japan to do research work. In Japan, he lived for 8 years, committed to making the world's first artificial brain. Successfully researching and developing the first machine: CBM (Cam Brain Machine) in order to make a human brain, he entered the Guinness Records, attracting attention from the world media. He had dreams of richness and opportunity of wealth, and hoped to be a millionaire. He invested USD100,000 in a private laboratory in Brussels, but ill-fated, high-tech bubbles burst, the investors no longer made high-tech risk investment and the laboratory went into bankruptcy. Out of principal, he lost USD100,000 and also lost his job. His research work did not go well. In Japan, he was pushed aside in a laboratory. In the USA, he had no funds to buy what he needs for the fifth CBM … In 2006, he came to China. He has been learning Chinese very hard, and has become a full-time professor at International Software College of Wuhan University.

According to Ms. Guo-qing LEI, Professor Garis's Chinese wife (the daughter of Zhen LEI, one of the first Republic Generals, now teaches also at Wuhan University), Garis has two great favors: reading and traveling. He read anytime, with five or six pens in his pockets, for fear of being out of ink when he is reading. To move from the United States to China, with his over ten thousand books, only the freight Professor Garis had to spend was nearly CNY150,000. Ms. LEI said: "We travel every weekends in China. During the winter and summer holidays, we travel abroad. This is our life." but "He is very monotonous, dozens of the same style and the same color clothes, and eats hamburgers and fast food all days. He does not care about other people. Once I was in the hospital to recuperate all day, When I came back, he asked me: 'my dear, where have you been?'" His brother is two years younger than him, who is a real estate tycoon in Australia and has private planes and private yachts, has said: "He does not understand the world". But his own evaluation is much richer: intelligence, curiosity, bookworm, travel, classical music, strongness and health. "I am a scientist and a research professor, a social activist, an author, someone said I am a thinker."

He was an artificial intelligence leader in the key laboratories in Utah, USA; in Brussels, Belgium; and in Tokyo, Japan etc.; he was responsible for completing two of the world's four "artificial brain", one in Japan and another in Belgium. In Japan, the intelligence of the artificial brain corresponds to that of a house cat. In 2000, Hugo obtained USD1,000,000 as research funds from the Brussels Government, and he returned to Brussels, Belgium to make an artificial brain. He controlled hundreds of robots that can behave themselves. For 1987–1991, he had learned from Professor Rrszard Michalski, a "machine learning" ancestor. He was a researcher at the Artificial Intelligence Center in George Mason University (GMU). In 1992, he obtained a doctorate in that University. He ranked in the top 10 of the 60 selected Ph.D. candidates in Europe with the most achievements in Natural Sciences and Engineering.

Professor Hugo has said: "Today's artificial intelligence machine is not what we think of—the human beings give them some functions and they will have, otherwise they will not have. They have produced junior evolution on their own in the nerves and smart, and would be evolved at an exponential speed. They would one day suddenly reach a status called 'singularity'. The idea is, when an artificial intelligence machine should reach a certain level of intelligence (singularity), they would be 'out of control' and accelerate their own evolution at a rate we could not imagine, so that it would overpass us very quickly and get far away ahead of us. What would happen to human beings at that time? What should we do? Is it already too late to consider it now?", "If you think these omnipotent artificial intelligence machines as Gods could not be made soon, then, in retrospect, people regarded Szilard as a madman when he forecast in the 1930s that a bomb (atom bomb) could destroy a city."

The United States' ABC latest film "Last Days on Earth", invited seven of the world's most famous scientists including Mr Hugo de Garis and Hawking to predict the end of Earth, in which Hugo's intelligent machines and Hawking's black hole etc. have been called the seven deadly threats that human would face. As the intelligent machines predicted by Hugo would come out in the next 50 years or so, it is the most urgent!

Below is an exclusive interview by a reporter.

Q: Why did you choose this research work?

A: The curiosity to the human brain and the human brain's imagination. A human being is only a machine made out of molecules, like computer chips, as well as programs. On the other hand, biological people will die and disappear, but artificial intelligence machines will not. So, this study is just as you are creating the Deity. Man makes technical progress, and makes the robots' emergence and development with artificial intelligence. But at last, the artificial intelligence machine would come true someday and achieve self-evolution. When such technologies reach a singularity, it does not need human to push off. The artificial intelligence machines much smarter than man would be produced in a short time, in years.

Q: Predict the future prospects of artificial intelligence machines?

A: In the next 20 years, they would be likely to appear in our homes. They would clean the houses for us, take care of the children, and talk to us, teach us the infinite knowledge with our human knowledge base on Earth. We would also be able to have sex with them, educated, and entertained and made laughing etc. by them. After 20 years, the worldwide brain manufacturing could create trillions of dollars in value every year. The artificial intelligence machines would be a trillion times smarter than we could be. No exaggeration to say, that the artificial intelligence machines communicate with human beings is just as hard as human beings try to communicate with rock. However, the true artificial intelligence would not be made within 30–40 years after my death. I could not live to see the real results of this work, which has been a source to make me frustrated and disappointed.

Q: How do you evaluate your research work?

A: Actually, I'm building the same thing that would become a God in the next few decades (though I may not live to see it completed). The expectation to manufacture a God has brought me a sense of religious awe. The feeling into my soul has strongly inspired me to continue this work. I seem to be schizophrenic. On one hand, I really want to make an artificial brain, and try to create it as clever as possible. I regard this as a great goal of human beings. On the other, I'm still worrying and shivering about the successful prospects of the artificial brain. The "high intelligent machine" is a trillion times artificial brain, which is really God stuff.

Q: You mentioned the concepts of Universism and Earthism in your book, *A Brief History of Intelligence*. Are you a Universist?

A: In the 21st century, it will be a dominant issue: whether human beings should produce the artificial intelligence machines that would be a trillion times more intelligent than human intelligence. I predict that human beings would be divided into two main political blocs: Universists and Earthists. To Universists, the manufacture of artificial intelligence is like religion, and is the mission of human beings; and to Earthists, they think it would be an adventure, and would lead to the destruction of human beings. So far, I'm only one who has predicted that the super artificial intelligence would bring war. I'm not 100% of a Universist. If I were, I would quietly engage in my research of artificial brain manufacturing, and would not go to warn the public against the artificial intelligence problems. When I should be dying, I would be proud of the "Father of Artificial Brain" called by people, but if history condemned me as "Father of Mass Death", this prospect would really make me scared. So I cannot do so. Open to the public, I am an Earthist. I am trying to warn everyone. I wrote *A Brief History of Intelligence: Who Will Replace Man as a Dominant Species*. Privately, I am a Universist, and I am cheering for my research. I thought I was making God, could not let me get excited.

Q: Some people say you are an epoch-making genius?

A: I'm not a genius, and I'm Professor. I like my job … I'm smart enough to read the works of genius, I'm smart enough to understand the true genius' what-to-do with their minds. What have they invented? They have a perfect system, and a genius creates a subject. Perhaps there are only 200 geniuses in the world. I cannot create a system as great as they do.

In the eyes of peers, Hugo is between "a complete freak" and "a beyond epoch genius", while the more common name is from the media: "Father of Artificial Brain", and "Hawking in the Field of Artificial Intelligence".

—Excerpted from Baidu

Summary

I. The Knowledge Structures of This Chapter

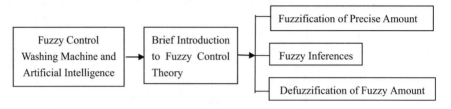

II. Review and Reflection

1. Recalling the automatic control, it leads to fuzzy control.
2. The problems that a fuzzy washing machine need resolve to determine many, a few, cleanness, dirtiness and so on, is actually a series of fuzzy problems.
3. Simulating human thinking by the three control rules of IF A, THEN B; IF A THEN B, ELSE C; and IF A AND B THEN C, and the fuzzy mathematical models.
4. The three operation steps of fuzzy controls.
5. From here to there, the washing machine seems to think, which relates to artificial intelligence and its developing prospects.

Chapter 4
Fuzzy Statistics and Fuzzy Probability

In the objective world, randomness and fuzziness coexist inside many things. Although fuzzy mathematics and probability theory cannot replace each other, they can infiltrate into each other. The fuzziness is introduced into stochastic phenomena in this chapter. Simple fuzzy statistics and fuzzy probability will be discussed.

4.1 Random Phenomena and Fuzzy Phenomena

Certain Phenomena and Uncertain Phenomena

There are a variety of phenomena in nature and human activities, which can be classified as certain phenomena and uncertain phenomena according to determinacy or not. Uncertainty can be subdivided into random phenomena, fuzzy phenomena and other uncertain phenomena. So, the mathematical models tackling these phenomena are often referred to as a deterministic mathematical model, a stochastic mathematical model, and a fuzzy mathematical model and so on.

The so-called certain phenomena are that, under certain conditions, the phenomena will certainly emerge. For example, "Like charges repel"; "At a standard atmospheric pressure, water heated to 100 °C will boil"; "A particle makes equally accelerated motion under a constant force"; a commodity scarcity will make prices increase, and so on.

The so-called random phenomena are that, even if the conditions appear exactly the same, the result of their appearance in general are not the same, or not completely certain and uncertain predictions. For example, throwing a homogeneous symmetric coin, it could be faced up or back up; newborn infants may be a boy or a girl; a shooter shooting at a target at one or more fires could hit the target, or not; In the economic field, GDP values are not the same every year, whose exact value is unknown in advance, and so on.

The so-called fuzzy phenomena are that, something does not occur in one fixed way or more ways, so that it would be hard to determine its models to describe and class its boundaries. For example, for the rain, there are many types of rains between light and heavy ones, but there are no clear boundaries among heavy rain and drizzle etc. For random phenomena, they are caused mostly by objective factors. However, fuzzy concept itself has no clear denotation. It is difficult to determine whether an object is in line with this concept, making uncertain boundaries.

Random phenomena and fuzzy phenomena both have uncertainty, but equally there is essentially difference, as shown in Table 4.1.

Probability and Membership Degree

Randomness is represented by probability, i.e., the probability is depicting the number indicator of uncertain size of an incident. It shows the objective possibility of the incident, and is an objective description of random phenomena. The objective

Table 4.1 The main differences between random and fuzzy phenomena

Comparing project	Random phenomena	Fuzzy phenomena
Different meaning	Uncertainties of all kinds of possible phenomena, i.e., features of results appear or not	Uncertainties of generic boundaries and state recognition of phenomena, i.e., phenomena show intermediary characteristics in the transition boundaries of "yes-no" or "plausible" phenomena
Different causing reasons	Due to inadequate conditions, it fails to makes causality of conditions and events	Due to uncertainty of objective things' transition boundaries; a variety of factors are intertwined in complex systems to produce ambiguity; due to uncertainty of boundaries caused by the recognition subject's differences in character, occupation, age, education etc.
Different respecting rules	Following the generalized law of causality, i.e., probability law (because a, b to a probable extent) is the breaking to the ordinary law of causality (because a, b)	Following the generalized law of excluded middle, i.e. the law of membership (which belongs to A to a probable extent, in the intermediate status of "A or non-A") is the breaking to the common law of excluded middle (either A or non-A)
Different sets presentations	The set is clear, which has clear connotation and denotation	The event is identified, but due to the uncertainty of the set's connotation and denotation, it is uncertain whether the event could be attributed to the set or not
Different research methods	Research by probability theory and mathematical statistics, and the probability describes the uncertainty of the event size	The membership degree is used to describe uncertain number size of the elements related to the set

meaning of probability can be rendered by the frequency stability in random experiments. Probability theory is a branch of mathematics to study and process random phenomena.

Fuzziness is portrayed through membership degree, i.e. the membership degree is the quantitative indicator of uncertain size that the elements are related to the set. It represents the objective possibility of the membership relation that elements are related to the set, and means an objective description of fuzzy phenomena. The frequency stability in the fuzzy statistic tests can be assumed as the objective meaning of the membership degree. Fuzzy mathematics is a branch of mathematics dealing with fuzzy phenomena.

Randomness means a breaking of the law of causality. Fuzziness is a breaking of the law of excluded middle. The use of probability theory is to grasp the broad law of causality from the random—the law of probability; and fuzzy mathematics is to establish the generalized law of excluded middle from the fuzzy—the law of membership.

The Coexistence of Randomness and Fuzziness

In the objective world, inside many things, randomness and fuzziness are coexisting, i.e., when we observe and study natural phenomena, we would encounter a lot of fuzzy problems. At the same time, the studied phenomena would be accompanied by randomness. For example, how much groundwater could be taken as a sign of water bursting into mine in a study of the mine inrush water problem? The answer would vary depending on the people, place and mine. The connection is not clear, with typical ambiguity; where and how much would the mine inrush water happen? It has a great deal of randomness.

It is worth mentioning that there's a difference among randomness, fuzziness and chaos. Chaos in Nonlinear Science refers to a determined but unpredictable motion, a system described by a deterministic theory, whose behavior is characterized by uncertainty—unrepeatable, unpredictable, this is chaos. Its external appearance is similar to purely random motion, i.e., both are unpredictable. The random motion differs from the chaos motion determined in dynamics. Its unpredictability comes from the instability of the motion. Studies have shown that chaos is the inherent characteristic of nonlinear dynamical systems, which is a common phenomenon in the natural world. The climate change is a typical chaotic motion.

In theory, things are generally linked. The random set and its drop shadow link randomness with fuzziness. Therefore, we should both learn fuzzy mathematics by the difference from randomness, and know it from the links.

4.2 Fuzzy Statistics

In the application of fuzzy mathematics, first of all it should establish a membership function, which is the basis and key to use fuzzy mathematics theory to solve practical problems. Due to the limitation of understanding things, people could only establish an approximate membership function. Whether it satisfies the fact or not has directly effects on the application of the membership function.

At present, how to determine a membership function is still a problem. Specific issues need specific treatments. But a membership function in certain situations can be established by fuzzy statistics method—Test methods to fuzzy statistics.

4.2.1   Test Methods to Fuzzy Statistics

In order to illustrate what a fuzzy statistics is, we explain it with probability statistics.

Probability Statistics

In probability theory, the probability of events can be obtained through a large number of randomized tests.

The most basic requirement for randomized tests is: it must be determined whether an event occurs or not in every test.

The characteristic of randomized tests is: under the same group of conditions, for n tests, event A occurs in m tests. So the Event A's frequency $= \dfrac{A \text{ occurs in } m \text{ tests}}{\text{total tests } n}$.
When n increases to infinity, the frequency of event A always tends to be a stable number. This stable number is called the probability of event A.

Fuzzy Statistics Methods of Membership Degrees

When determining the membership degree of element u_0 to set A, it can use similarly the randomized tests method—the test methods to fuzzy statistics to solve the probability, i.e., through a large number of fuzzy statistics tests to find out the law of statistics.

The basic requirement of fuzzy statistics tests is: in every test, you have to make a definitive judgement whether u_0 belongs to A, i.e., in every test, A must be a definitive common set.

The characteristic of fuzzy statistics tests is: if in n tests, element u_0 belongs to given set A for m times, the membership frequency of u_0 to A is
frequency $= \dfrac{u_0 \text{ belongs to } A \text{ for } m \text{ times}}{\text{total tests } n}$. When n increases to infinity, the membership frequency of u_0 to $\underset{\sim}{A}$ always tends to be a stable number. This stable number is called the membership degree of u_0 to $\underset{\sim}{A}$.

To sum up the fuzzy statistics tests, there are usually four basics:
(1) The domain U; (2) A fixed element u_0 in the domain U; (3) The common Set A with variable boundaries in the domain U, it reflects the uncertainty of the denotation of fuzzy concept, i.e., boundaries are not clear; (4) The test conditions, which reflect the influence of fuzzy concept $\underset{\sim}{A}$ in a partition and restrict the motion of A.

It is not difficult to find that in each fuzzy statistical test, u_0 is fixed, A is variable. If you think the probability is to examine whether "some variable dots" falls into "an immobile circle", the fuzzy statistics can be compared to examine whether "a variable circle" covers "some immobile dots".

A case is introduced in Chap. 1. Professor Nan-lun ZHANG and other scholars in Wuhan Institute of Building Materials have made a sampling test for the fuzzy set of "youth". This is one of the typical fuzzy statistics tests. It is not difficult to solve that a 27-year-old person belongs to the "youth" with the membership degree $\mu_A(27) = 0.78$.

Fuzzy Statistics Methods of Membership Functions

The basic steps to use the fuzzy statistics method to construct the membership function curve are shown as follows:

Step 1: Grouping

In the fuzzy statistics test, each result of it is a subset of domain U. If domain U is the real axis, the statistical result is a real interval. When the number of fuzzy statistic tests is large, the fuzzy statistics intervals should be grouped.

Generally, the group number has to be determined according to the interval capacity of statistics data. The groups could not be too many or too few, at least 5 groups, or at most 20 groups. It is suitable that each group should have average 5 data or more. When the data are grouped, the statistics data interval should be arranged first by the endpoint value in size order, then the data interval should be divided into k non-intersectant grouping intervals (x_i, x_{i+1}). In general, the group distances are required equidistantly. The midpoint value of each group interval $\xi_i = \dfrac{x_i + x_{i+1}}{2}$, called the group median, which can be used to replace the average within group data. The specific grouping methods are shown as follows:

1. To determine the upper bound and the lower bound of the statistical range. Determination of the upper bound can be a little bit bigger than the maximum value at the right endpoint of the range while the lower bound can be a little bit smaller than the minimum value at the left endpoint of the range.
2. To determine the number of groups k and the grouping intervals' values, and set the group interval l to make $l = \dfrac{\text{upper bound} - \text{lower bound}}{k}$.
3. To make statistics to cover every grouping statistical data interval (x_i, x_{i+1})'s number n_i by calling it out one by one, called the coverage frequence.

Step 2: Listing a Statistical Table

After the interval grouping of statistical data, the statistical table is listed. It can compress a large amount of statistical data intervals into refined representation so as to supply data to make a membership function curve. The items of the statistical table include: Group No., Group Interval, Group Median, Coverage Frequence, Coverage Frequency and so on.

Coverage Frequence n_i refers to the number covering the ith grouping interval (x_i, x_{i+1}) of the statistical data interval. Coverage Frequency refers to the ratio of Coverage Frequence n_i of the ith grouping interval to the total number n of the statistical data interval, that is, $\mu_i = \dfrac{n_i}{n}(i = 1, 2, \ldots, k)$, the Coverage Frequency μ_i is also called membership frequency.

Step 3: Making a Membership Function Graph

Suppose that the abscissa as domain U axis, to the ordinate value representing the membership frequency μ. Set up the coordinate system. On the U axis, set the grouping's upper bound and lower bound, and mark out every Group Median in turn. Then take values on both equidistant sides by using every Group Median as the center to determine the location of grouping points. Draw each height dashed at every Group Median with each value of the coverage frequency μ_i. At last use a smooth curve to join up each vertex of the dasheds. This smooth curve is the sought membership function curve.

For making the membership function curve of "youth" by Nan-lun ZHANG and other scholars, try to take $k = 23$, and the group interval $l = \dfrac{\text{upper bound} - \text{lower bound}}{k} = \dfrac{36.5 - 13.5}{23} = 1$, i.e., calculating once across 1 age interval, and seeking out each Group Median. For example, the Group Median of $(13.5, 14.5)$ is $\xi_1 = \dfrac{13.5 + 14.5}{2} = 14$. Appearing twice means the coverage frequence $n_1 = 2$. The coverage frequency $\mu_1 = \dfrac{n_1}{n} = \dfrac{2}{129} = 0.0155$, so as to get the frequency statistics in Table 1.3 of Chap. 1 in this book. At last make the membership function curve of "youth" (see Fig. 1.15).

It is necessary to note that in fuzzy statistical tests for the fuzzy concept of human natural language, the principles to be followed are that the participants must be familiar with the concept of fuzzy words and have the ability to express this concept with the number approximately, requiring each participant to express himself with an exact set of the fuzzy concept as closely as possible.

4.2.2 Fuzzy Data and Operations

Let us understand a few examples.

A high school student has sports for 1–2 h at a time, 2–4 times a week. How many hours does he practices a week? For another example, someone wants to go out traveling for 7–8 days and suffers from stomachaches. How many tablets should he carry with him if the medicines for stomachaches need to be taken 5–7 tablets for a dosage, 3–4 times a day for adults? Traditional mathematics is difficult to deal with such problems, but most people can make a ballpark estimate of them.

The Fuzzy Number

Definition 4.1 Suppose U is a domain, $\{A_1, A_2, \ldots, A_n\}$ is a subset of U, and u represents a real-valued function u corresponding to $[0, 1]$: $U \to [0, 1]$. If there is a statement X in domain U, the membership function of its subset is represented as $\{\mu_1(X), \mu_2(X), \ldots, \mu_n(X)\}$. In the discrete case, statement X's fuzzy number can be expressed as follows:

$$\mu_U(X) = \frac{\mu_1(X)}{A_1} + \frac{\mu_2(X)}{A_2} + \cdots + \frac{\mu_n(X)}{A_n}, \tag{4.1}$$

where $+$ represents the interface symbol, it does not mean an addition, $\frac{\mu_i(X)}{A_i}$ represents the extent that statement X belongs to subset A_i.

When U is continuous, statement X's fuzzy number can be expressed as:

$$\mu(X) = \int_{x \in X} \frac{\mu_i(X)}{A_i} \tag{4.2}$$

Example 4.1 Assume that X represents some high school student who has sports for a few hours a day, denoted by the fuzzy number $\mu_U(X)$. Domain U can be viewed as the integer field, that is, the exercise hours. Suppose that $U = \{0, 1, 2, 3, 4, 5\}$, the membership function that X represents the fuzzy numbers of sports' hours a day can be expressed as:

$$\{\mu_0(X) = 0.25, \mu_1(X) = 0.4, \mu_2(X) = 0.2, \mu_3(X) = 0.1, \mu_4(X) = 0.05, \mu_5(X) = 0\},$$

then the fuzzy number of X can be expressed as:

$$\mu_U(X) = \frac{0.25}{0} + \frac{0.4}{1} + \frac{0.2}{2} + \frac{0.1}{3} + \frac{0.05}{4} + \frac{0}{5}.$$

To use the same method in Example 4.1, it is not difficult to solve the problem introduced at the beginning of this section—the problem of carrying the tablets for stomachaches. You may wish to have a try.

Example 4.2 Assume that X represents the city residents' feeling to "The Industrial Park's sewerage" along the Songhua River, expressed with the fuzzy number $\mu_U(X)$, Demain U is,

$$U = \{1 = \text{very bad}, 2 = \text{bad}, 3 = \text{common}, 4 = \text{slight}, 5 = \text{uninfluential}\}.$$

If the membership function of X's feeling is

$$\{\mu_1(X) = 0.25, \mu_2(X) = 0.6, \mu_3(X) = 0.1, \mu_4(X) = 0.05, \mu_2(X) = 0\},$$

the fuzzy number can also be expressed as

$$\left\{\mu_U(X) = \frac{0.25}{1} + \frac{0.6}{2} + \frac{0.1}{3} + \frac{0.05}{4} + \frac{0}{5}\right\}.$$

Example 4.3 Assume that the fuzzy number $\mu_U(X)$ represents the return on investment of X Company. Domain U can be viewed as a real number field, that is, the return on investment. Suppose that $U = \{2\%, 5\%, 10\%, 20\%\}$, such that the membership function of the fuzzy number for X Company's return on investment is

$$\{\mu_2(X) = 0.2, \mu_5(X) = 0.4, \mu_{10}(X) = 0.3, \mu_{20}(X) = 0.1\}.$$

Then the fuzzy number for X Company's return on investment can be expressed as

$$\mu_U(X) = \frac{0.2}{2\%} + \frac{0.4}{5\%} + \frac{0.3}{10\%} + \frac{0.1}{20\%}.$$

Definition 4.2 Assume that x represents the discrete fuzzy number. The language variable $\{L_i | i = 1, 2, \ldots, k\}$ for the ordered sequence in Domain U, such that $\mu_{L_i}(x) = m_i$ is the membership degree of the fuzzy number x to the language variable L_i, $\sum_{i=1}^{n} \mu_{L_i}(x) = 1$, then $x_f = \sum_{i=1}^{n} m_i L_i$ is called the defuzzification value of the fuzzy number x.

4.2.3   Descriptive Statistics of Fuzzy Data

In the traditional statistical inference methods, in order to understand the unknown population parameters, we often resort to some of the evaluation criteria, and identify appropriate statistics to estimate population parameters. The average is one of the most important population parameters to understand the population con-centration trend. We often use the unbiased estimator, such as the sample mean to estimate the population concentration trend. However, in everyday life, the popu-lation mean is often a fuzzy, uncertainty, semantic variable or a possible range, so that the traditional evaluation estimator, criteria and methods could not apply to this situation. So, we have to learn some of the basic descriptive statistics for the fuzzy data—the fuzzy sample mean, the fuzzy sample mode, and the fuzzy sample median etc.

The Sample Mean of Fuzzy Data

Definition 4.3 The Discrete Fuzzy Sample Mean

Assume that U represents a domain, $L = \{L_1, L_2, \ldots, L_k\}$ are k language variables in domain U, $\left\{x_i = \frac{m_{i1}}{L_1} + \frac{m_{i2}}{L_2} + \cdots + \frac{m_{ik}}{L_k}, i = 1, 2, \ldots, n\right\}$ is for a group of fuzzy samples, such that $\sum_{j=1}^{k} m_{ij} = 1$, then the fuzzy sample mean is defined as:

$$F\bar{s} = \frac{\frac{1}{n}\sum_{i=1}^{n} m_{i1}}{L_1} + \frac{\frac{1}{n}\sum_{i=1}^{n} m_{i2}}{L_2} + \cdots + \frac{\frac{1}{n}\sum_{i=1}^{n} m_{ik}}{L_k}, \tag{4.3}$$

in which, m_{ij} is the membership degree of the ith sample relative to the language variable L_j.

Example 4.4 The discrete-type fuzzy sample mean is used for commodity satisfaction surveys.

A new commodity came into the market. The manufacturer intended to explore the consumers' satisfaction to it and invited 5 consumers: A, B, C, D, and E in the streets to make a survey. The membership degree of every consumer's commodity satisfaction is shown in Table 4.2.

Table 4.2 The membership options of 5 consumers' satisfaction to the commodity

Satisfaction level	L_1 very dissatisfied	L_2 dissatisfied	L_3 common	L_4 satisfied	L_5 very satisfied
A	0	0.5	0.5	0	0
B	0	0.5	0.8	0.2	0
C	0	0.3	0.7	0	0
D	0	0	0	0.9	0.1
E	0	0	0.2	0.8	0

Then the fuzzy sample mean is:

$$F\bar{s} = \underbrace{\frac{\frac{1}{5}(0+0+0+0+0)}{\text{very dissatisfied}}}_{} + \underbrace{\frac{\frac{1}{5}(0.5+0+0.3+0+0)}{\text{dissatisfied}}}_{} + \underbrace{\frac{\frac{1}{5}(0.5+0.8+0.7+0+0.2)}{\text{common}}}_{}$$

$$+ \underbrace{\frac{\frac{1}{5}(0+0.2+0+0.9+0.8)}{\text{satisfied}}}_{} + \underbrace{\frac{\frac{1}{5}(0+0+0+0+0.1)}{\text{very satisfied}}}_{}$$

$$+ \frac{0}{\text{very dissatisfied}} + \frac{0.16}{\text{dissatisfied}} + \frac{0.44}{\text{common}} + \frac{0.38}{\text{satisfied}} + \frac{0.02}{\text{very satisfied}}.$$

This fuzzy sample mean represents the meaning that: the membership degree of "very satisfied" is 0.02; the membership degree of "satisfied" is 0.38; the membership degree of "common" is 0.44; the membership degree of "dissatisfied" is 0.16; the membership degree of "very dissatisfied" is 0. This fuzzy mean is a fuzzy number, showing that the average satisfaction of this commodity is "common" at most, and the next is "satisfied".

Definition 4.4 The Continuous Fuzzy Sample Mean (The samples as a continuous and uniform distribution)

Assume that U represents a domain, $L = \{L_1, L_2, \ldots, L_k\}$ are k language variables in domain U, $\{x_i = [a_i, b_i], i = 1, 2, \ldots, n\}$ is for a group of fuzzy samples in domain U, then the fuzzy sample mean is defined as:

$$F\bar{s} = \left[\frac{1}{n} \sum_{i=1}^{n} a_i, \frac{1}{n} \sum_{i=1}^{n} b_i \right]. \tag{4.4}$$

Example 4.5 In a university, some students of 2011 were looking for a job. A survey showed that 5 of the students from mathematics and information science college expected their salary to be a group of fuzzy samples for [CNY20,000; CNY30,000], [CNY30,000; CNY40,000], [CNY40,000; CNY60,000], [CNY50,000; CNY80,000], [CNY40,000; CNY70,000], then according to Definition 4.4, its fuzzy sample mean is:

$$F\bar{s} = \left[\frac{2+3+4+5+4}{5}, \frac{3+4+6+8+7}{5} \right] = [3.6, 5.6] \text{ (Ten thousand yuan)}.$$

This information could provide the companies for references, which need university students, in order to understand the present salary that the students would want at Mathematics and Information Science College.

The Sample Mode of Fuzzy Data

Definition 4.5 The Discrete Fuzzy Sample Mode

Assume that U represents a domain, $L = \{L_1, L_2, \ldots, L_k\}$ are k language variables in domain U, $\left\{ x_i = \frac{m_{i1}}{L_1} + \frac{m_{i2}}{L_2} + \cdots + \frac{m_{ik}}{L_k}, i = 1, 2, \ldots n \right\}$ is for a group of fuzzy samples, such that $\sum_{j=1}^{k} m_{ij} = 1$, suppose $T_j = \sum_{j=1}^{m} m_{ij}$, then the L_j that has the maximum value of T_j is the fuzzy sample mode, denoted as:

$$Fmo = \left\{ L_j | \text{relative to } j \text{ item, to make } T_j = \max_{j=1,2,\ldots,k} T_j \right\}. \tag{4.5}$$

If there are two or more sets of L_j with the same maximum values T_j, then this group of data is said to have more than one mode or multiple consensus.

Example 4.6 The test paper database is preestimated for the entrance examination of senior middle schools in a city some year. In process of preparing test questions, some subject experts and educational test experts are invited according to their professions to preestimate the topics and the students about the membership degrees of difficulty $p(0 \leq p \leq 1)$. The 12 experts denoted by $S_1, S_2, \ldots, S_{12}$ are selected to preestimate the difficult membership degrees of some test paper, with their professions and experiences. Use $\{L_1, L_2, L_3, L_4, L_5\}$ to denote respectively {very difficult, difficult, medium, easy, very easy}. The membership degrees of difficulty that the 12 experts have predicted on the test questions are shown in Table 4.3.

If the traditional 5-level survey is used, the prediction is found as shown in Table 4.4.

By using a traditional mode method, a forecast result of difficult levels to the test questions is "easy", which came from the highest votes by the educational testing experts (S_2, S_7, S_8, S_{12}: four votes). But among 12 experts, only 4 of them have predicted the result. It does not seem to make much sense. If calculated with the

Table 4.3 The membership degrees of difficult levels of the predicted test questions

Experts/difficult levels	L_1 very difficult	L_2 difficult	L_3 medium	L_4 easy	L_5 very easy
S_1	0.5	0.4	0.1		
S_2			0.4	0.6	
S_3		0.6	0.4		
S_4		0.4	0.6		
S_5	0.4	0.6			
S_6				0.1	0.9
S_7			0.4	0.6	
S_8			0.4	0.6	
S_9		0.4	0.6		
S_{10}			0.1	0.1	0.8
S_{11}	0.4	0.6			
S_{12}			0.4	0.6	
Total	1.5	2.8	3.4	2.6	1.7

Table 4.4 The prediction of difficult levels of the test questions

Experts/difficult levels	L_1 very difficult	L_2 difficult	L_3 medium	L_4 easy	L_5 very easy
S_1	•				
S_2				•	
S_3		•			
S_4			•		
S_5		•			
S_6					•
S_7				•	
S_8				•	
S_9			•		
S_{10}					•
S_{11}	•				
S_{12}				•	
Total	2	2	2	4	2

fuzzy membership degrees, then the question difficulty is predicted as "medium", whose total membership degree of 3.4 is much greater than that of "easy" with 2.6. In contrast, the fuzzy sample mode would be more objective than the traditional mode, and also could achieve consensus on this forecast for the question difficulty.

The Sample Median of Fuzzy Data

Definition 4.6 The Discrete Fuzzy Sample Median

Assume that U represents a domain, $L = \{L_1, L_2, \ldots, L_k\}$ are k ordered variables in domain U, $\left\{x_i = \frac{m_{i1}}{L_1} + \frac{m_{i2}}{L_2} + \cdots + \frac{m_{ik}}{L_k}, i = 1, 2, \ldots, n\right\}$ is for a group of fuzzy samples drawn from domain U, such that $\sum_{j=1}^{k} m_{ij} = 1$, x_{if} is called the defuzzification value corresponding to the fuzzy samples x_i.

Let $x_{(i)}$ for the ordered sample values after x_i is ordered. The discrete fuzzy sample median is defined as:

$$Fme\ (X) = \begin{cases} x_{\left(\frac{n+1}{2}\right)}, & \text{if } n \text{ is an odd number;} \\ \dfrac{x_{\left(\frac{n}{2}\right)} + x_{\left(\frac{n}{2}+1\right)}}{2}, & \text{if } n \text{ is an even number.} \end{cases} \tag{4.6}$$

Example 4.7 The discrete fuzzy sample median is applied to the product pricing survey.

A company is going to develop a new beverage and study out the price list. Assume that the beverage has 5 proposed prices, such as 20, 25, 30, 35, and 40 (unit: CNY). The company wants to explore what kind of prices could be accepted by consumers, and has selected randomly 6 consumers for the market researches. Each consumer is required to comment on every price's membership degree. The result is shown in Table 4.5.

As the number of samples $n = 6$ is an even number, $x_{(3)f} = 25 \times 0.6 + 30 \times 0.4 = 27$. Similarly, $x_{(4)f} = 31$, and the sample values corresponding to $x_{(3)f}$, $x_{(4)f}$ are:

$$x_{(3)} = x_3 = \frac{0}{20} + \frac{0.6}{25} + \frac{0.4}{30} + \frac{0}{35} + \frac{0}{40},$$

$$x_{(4)} = x_6 = \frac{0}{20} + \frac{0}{25} + \frac{0.8}{30} + \frac{0.2}{35} + \frac{0}{40}.$$

Table 4.5 The 6 participants' membership selection from 5 prices

Price	20	25	30	35	40	x_{if}
1	0	0	0.5	0.5		32.5
2	1	0	0	0	0	20
3	0	0.6	0.4	0	0	27
4	0	0.7	0.3	0	0	26.5
5	0	0	0.4	0.3	0.3	34.5
6	0	0	0.8	0.2	0	31

Thus, the fuzzy sample median is:

$$Fme(X) = \frac{0}{20} + \frac{\frac{0.6+0}{2}}{25} + \frac{\frac{0.4+0.8}{2}}{30} + \frac{\frac{0+0.2}{2}}{35} + \frac{0}{40}$$

$$= \frac{0}{20} + \frac{0.3}{25} + \frac{0.6}{30} + \frac{0.1}{35} + \frac{0}{40}.$$

Seen from the sorting of membership, the result out of the fuzzy sample median is 30 yuan (CNY), i.e., the price of 30 yuan (CNY) could be accepted to a higher extent by the general public.

As we all know, the mode, the mean and the median are three kinds of common measurement of central tendency statistics. Their solution is simple and clear to express the datum's characteristic and the conveyed information. Therefore, it is a useful and important way to use rationally the fuzzy samples' mean, mode, and median etc. in the statistic studies and analyses of practical problems, based on the human fuzzy complex thinking.

4.3 Fuzzy Probability

In Sect. 4.1 of this chapter, we have discussed the relationship and difference between randomness and fuzziness, and illustrated that the fuzziness and the randomness are interactive and coexistent inside many things in the real world. This section will introduce the fuzziness into the random phenomena, and discuss the simple fuzzy probability.

4.3.1 Fuzzy Probability Theory

In daily life, we often encounter this kind of problems—"How often will the probability (possibility) of good weather happen tomorrow?" Obviously, that "Tomorrow will be a good day" is a random event, which is a probability problem. However, it is difficult to establish a clear set of criteria to judge "a good day", which has a great deal of ambiguity. Easy to see that such questions are beyond the scope of classical probability theory. They are no longer simple fuzzy problems or probability problems, but the fuzzy probability problems that we are going to study.

In general, the scope within which the fuzzy probability theory discusses is often the fuzzy event's probability, the event's fuzzy probability and the fuzzy event's fuzzy probability.

The fuzzy event's probability is defined as the fact that the event itself is fuzzy, but the probability is an ordinary numerical value. As the above example, "How often will the probability (possibility) of good weather happen tomorrow?"; "How high will the probability of the next match win excitedly?"; "How many shoots can be the probability to hit the target?"; "What is the probability of high quality chalk when a piece of chalk is taken randomly from this chalk box?" and so on. These are some problems of the fuzzy event's probability.

The event's fuzzy probability refers to the fact that the event itself is explicit, but the probability is fuzzy. For example, there is a match between two teams. How high could be the probability that Team A would win? Then, it would be hard to give an exact numerical answer because we might usually have a habit of estimating a team with the words: "very high", "very low", "very tiny" or such fuzzy concepts to describe it. We would be unwilling to do so with a precise number or an interval although "Team A has won" usually could be a common event. If we estimated a team with some fuzzy concepts, it would be suitable and flexible to do so despite the loss of accuracy. It is necessary to represent the probability with a precise number, but in some occasions, it needs fuzziness. We would rather say: "'Team A will win' has 'very high' probability" than "'Team A will win' has the probability of 0.89". It is called language probability by Professor L. A. Zadeh to use these fuzzy words to represent the possibility of an event. Thus, the event's fuzzy probability is also known as the Language Probability of a common event.

The fuzzy event's fuzzy probability is called the fuzzy event's language probability, which refers to the fact that both the event and the probability are fuzzy. In daily life, it is very normal that people would talk about something with the fuzzy event's language probability. For example, "It is possible that tomorrow will make good weather."; "The person you are looking for may be very tall."; "It is possible that most people would embrace this solution." and so on. Nay, even some propositions to use vague language would help to understand although it could be imprecise. As in the classical probability theory, the Chebyshev big-data theorem can be said to be that "in repeated independent trials, it could be very tiny that the large deviations would take place in the probability and frequency of the event."

This section only describes the fuzzy event's probability (limited discrete random variables).

4.3.2 Probability of Fuzzy Events

To the fuzzy event that "tomorrow may rain", how could we measure its possibility? This is a problem to compute the fuzzy event's probability. Here, on the basis of classical probability theory, the fuzzy event's probability is studied.

Ordinary Events and Their Probability

In the classical probability theory, every possible result of a random test E is called a basic event ω, such that the set made up of all events is called the sample space, denoted by U.

Some set A of the basic event ω is called an event, which is a subset A of the sample space. The sample space can be regarded as an event, known as the inevitable event, such that an empty set $\varnothing$ denotes an impossible event.

Definition 4.7 Assume that the sample space $U = \{\omega|\omega$ is a basic event$\}$ of random tests E, A is an event, such that the real number $P(A)$ is called the probability of event A, if it meets the following three conditions:

(1) For each event A, there is $0 \leq P(A) \leq 1$;
(2) For an inevitable event U, there is $P(A) = 1$;
(3) Suppose that the event $A_i(i = 1, 2, \ldots$ is not compatible with each other, i.e., $A_i \cap A_j = \varnothing(i \neq j)$, then:

$$P\left(\bigcup_{i=}^{\infty} A_i\right) = \sum_{i=1}^{\infty} P(A_i).$$

If a sample space U is a finite set, that is, $U = \{\omega_1, \omega_2, \ldots, \omega_n\}$, such that the probability $P(\omega_i) = p_i(i = 1, 2, \ldots, n)$ of an basic event ω_i, then the probability of event A is:

$$P(A) = \sum_{\omega_i \in A} p_i = \sum_{i=1}^{n} x_A(\omega_i)p_i,$$

where $x_A(\omega_i)$ is the characteristic function of event A.

Similarly with the fuzzy set theory, the ordinary event and its probability as defined above can be expanded to the fuzzy event.

Fuzzy Events and Their Probability

Assume that a sample space is U, if a fuzzy subset A of U is a random variable, then A is called a fuzzy event, such that the fuzzy event A is a fuzzy subset of the sample space U.

For example, suppose that a product's sample space $U = \{0, 1, 2, \ldots, 100\}$ with some waste products. We take 100 samples randomly from the products. A denotes "few waste products in the samples" and B indicates "about 4 waste products in the samples", then, A and B are both fuzzy events. They can be represented by the following fuzzy subsets:

$$A = \frac{1}{0} + \frac{1}{1} + \frac{0.8}{2} + \frac{0.5}{3},$$
$$B = \frac{0.6}{2} + \frac{1}{3} + \frac{1}{4} + \frac{1}{5} + \frac{0.6}{6}.$$

Definition 4.8 If a sample space U is a finite set, that is, $U = \{\omega_1, \omega_2, \ldots, \omega_n\}$, such that the probability $P(\omega_i) = p_i(i = 1, 2, \ldots, n)$ of an basic event ω_i; the membership function of fuzzy event $\underset{\sim}{A}$ is $\mu_{\underset{\sim}{A}}(\omega_i)$, then the fuzzy event A's probability is defined as:

$$P(\underset{\sim}{A}) = \sum_{i=1}^{n} \mu_{\underset{\sim}{A}}(\omega_i)p_i. \tag{4.7}$$

Example 4.8 For a census of certain region, an incidence of chronic bronchitis is shown in Table 4.6.

Table 4.6 A regional incidence of chronic tracheitis

Age group (years old)	Incidence of the disease (%)
Below 15	2.8
15–25	3.7
25–35	5.0
35–45	7.6
45–55	11.0
Above 55	14.2

Suppose that "the young" $= \dfrac{1}{(15-25)} + \dfrac{0.6}{(25-35)}$, and "the old" $= \dfrac{1}{(\text{above } 55)} + \dfrac{0.3}{(45-55)}$. Solve the probability of chronic tracheitis for "the young" and "the old" in this region.

Solution It is clear that "the young" and "the old" are fuzzy events, then:
The probability that "the young" suffer from chronic tracheitis is $0.37 + 0.6 \times 0.05 = 0.067$, while the probability that "the old" suffer from chronic tracheitis is $0.142 + 0.3 \times 0.11 = 0.173$.

Example 4.9 The sales quantity of a cellphone monopoly sales counter is random every day. According to the statistics, we know that the probability distribution of the sold cellphones is shown in Table 4.7, ask: What is the probability that about 8 cellphones have been sold every day?

Table 4.7 The probability distribution of the cellphones sold every day

Sales quantity	0	1	2	3	4	5	6	7	8	9	10	11 ...
Probability $P(\omega_i)$	0.01	0.02	0.02	0.03	0.04	0.04	0.05	0.1	0.15	0.2	0.15	0.1 ...

Solution Suppose A represents the event of "about 8 cellphones sold", then A is a fuzzy event. The everyday sales quantity of cellphones $U = \{1, 2, 3, 4, \ldots\}$, such that

$$A = \frac{0.4}{6} + \frac{0.7}{7} + \frac{1}{8} + \frac{0.7}{9} + \frac{0.4}{10},$$

then the probability of A is

$$P(A) = \sum_{i=1}^{n} \mu_A(\omega_i) p_i$$
$$= 0.05 \times 0.4 + 0.1 \times 0.7 + 0.15 \times 1 + 0.2 \times 0.7 + 0.15 \times 0.4 = 0.44.$$

Example 4.10 Shoot at the target until it is hit. Suppose that every shoot is independent from each other, the probability of hitting the target every time is p; A represents the event that "it hits the target after shooting only several times". Suppose $A = \frac{1}{1} + \frac{0.8}{2} + \frac{0.6}{3} + \frac{0.4}{4}$, solve the probability of A.

Solution It is clear that A is a fuzzy event. Take the firing times as needed to hit the target for Domain $U = \{0, 1, 2, 3, 4, \ldots\}$, the probability of the independent incident of "hitting the target at the ith time" is $p_i = (1 - p)^{i-1} p$, so the probability of A occurrence is:

$$P(A) = \sum_{i=1}^{n} \mu_A(\omega_i) p_i = \sum_{i=1}^{n} \mu_A(\omega_i)(1 - p)^{i-1} p$$
$$= p + 0.8(1 - p)p + 0.6(1 - p)^2 p + 0.4(1 - p)^3 p.$$

Example 4.11 Given that the reject rate of certain product is 0.01, take 100 randomly, A represents the event of "few rejects in the sampling products", and B the event of "about 4 rejects in the sampling products", taking the number of waste products as Domain $U = \{0, 1, 2, 3, 4, \ldots, 100\}$, and suppose $A = \frac{1}{0} + \frac{1}{1} + \frac{0.8}{2} + \frac{0.5}{3}$, $B = \frac{0.6}{2} + \frac{1}{3} + \frac{1}{4} + \frac{1}{5} + \frac{0.6}{6}$. Solve the probability of A and B.

Solution Obviously, A and B are fuzzy events. The probability that there are i waste products out of 100 samples is:

$$p_i = C_{100}^i \times 0.01^i \times (1 - 0.01)^{100-i} (i = 1, 2, \ldots, 100).$$

So, the probability of A is:

$$P(A) = \sum_{i=1}^{n} \mu_{\underset{\sim}{A}}(\omega_i)p_i = \mu_{\underset{\sim}{A}}(0)p_0 + \mu_{\underset{\sim}{A}}(1)p_1 + \mu_{\underset{\sim}{A}}(2)p_2 + \mu_{\underset{\sim}{A}}(3)p_3$$

$$= 0.37 + 0.37 + 0.8 \times 0.18 + 0.5 \times 0.06 = 0.91.$$

The probability of B is:

$$P(B) = \sum_{i=1}^{n} \mu_{\underset{\sim}{B}}(\omega_i)p_i = \mu_{\underset{\sim}{B}}(2)p_2 + \mu_{\underset{\sim}{B}}(3)p_3 + \mu_{\underset{\sim}{B}}(4)p_4 + \mu_{\underset{\sim}{B}}(5)p_5 + \mu_{\underset{\sim}{B}}(6)p_6$$

$$= 0.6 \times 0.18 + 0.06 + 0.02 + 0.003 + 0.6 \times 0.0005 = 0.2.$$

From the above simple examples, we can see that the fuzzy probability calculation is actually for averaging, or for the mathematical expectation of membership functions. When the membership function of an event is given, the problem is not difficult to solve.

According to the definition of the fuzzy event's probability, many results in the classical probability theory can be extended to the fuzzy probability theory. For the properties of fuzzy probability, several common conclusions are given below.

The Basic Properties of Fuzzy Event's Probability

Suppose that the sample space is U, A and B are fuzzy events. Then, according to the operation property of fuzzy sets, we have:

(1) If $A \subseteq B$, then

$$P(A) \le P(B); \tag{4.8}$$

(2) For any fuzzy event

$$A, P(A^c) = 1 - P(A); \tag{4.9}$$

(3) $$P(A \cup B) = P(A) + P(B) - P(A \cap B). \tag{4.10}$$

Example 4.12 In a game of playing at dice, the basic events that may arise as 1, 2, 3, 4, 5, 6. So domain $U = \{1, 2, 3, 4, 5, 6\}$. A represents the fuzzy event of "big number", B the fuzzy event of "small number", and C the fuzzy event of "both big number and small number", such that $A = \dfrac{0.2}{3} + \dfrac{0.4}{4} + \dfrac{0.8}{5} + \dfrac{1}{6}$, $B = \dfrac{1}{1} + \dfrac{0.8}{2} + \dfrac{0.4}{3} + \dfrac{0.2}{4}$, $C = A \cap B = \dfrac{0.2}{3} + \dfrac{0.2}{4}$. Solve the following fuzzy event's probability:

(1) "no big number" (recorded as Event D);

(2) "either big number or small number" (recorded as Event E).

Solution According to the meaning of the question, the probability of the fuzzy events $A, B, C = A \cap B$ are in turn:

$$P(A) = \sum_{i=1}^{6} \mu_A(\omega_i)p_i = \sum_{i=1}^{6} \mu_A(i)p_i,$$

$$P(B) = \sum_{i=1}^{6} \mu_B(\omega_i)p_i = \sum_{i=1}^{6} \mu_B(i)p_i,$$

$$P(C) = \sum_{i=1}^{6} \mu_C(\omega_i)p_i = \sum_{i=1}^{6} \mu_C(i)p_i.$$

And the probability of throwing a dice to get i points is $\frac{1}{6}$, so:

$$P(A) = 0.2 \times \frac{1}{6} + 0.4 \times \frac{1}{6} + 0.8 \times \frac{1}{6} + 1 \times \frac{1}{6} = 0.4,$$

$$P(B) = 1 \times \frac{1}{6} + 0.8 \times \frac{1}{6} + 0.4 \times \frac{1}{6} + 0.2 \times \frac{1}{6} = 0.4,$$

$$P(C) = 0.2 \times \frac{1}{6} + 0.2 \times \frac{1}{6} = 0.067.$$

So from the formula (4.9), we get $P(D) = P(A^c) = 1 - 0.4 = 0.6$, from the formula (4.10), we have:

$$P(E) = P(A \cup B) = P(A) + P(B) - P(A \cap B) = 0.4 + 0.4 - 0.067 = 0.733.$$

Exercise 4

1. Determining the membership function curve for "tall Chinese" by the fuzzy statistics methods, and seeking the membership degree for the Chinese with height of 175 cm.
2. Assuming that a company has a 10-person plan to spend the weekends in going outing, there are 4 select travel sites: Xiachuan Island, Botanical Gardens, Maofeng Mountain and Baishui Village. Table 4.8 represents the voted results of the questionnaire with the fuzzy membership degree and the traditional polling.

   (1) Use the principle of the fuzzy sample mean, choose reasonably a travel destination, and give the reasons;

Table 4.8 The questionnaire results of select travel sites

Vote	Xiachuan Island	Botanical Gardens	Maofeng Mountain	Baishui Village	Xiachuan Island	Botanical Gardens	Maofeng Mountain	Baishui Village
S_1	0.4	0.6	0	0		●		
S_2	0.5	0	0.4	0.1	●			
S_3	0.1	0	0.4	0.5				●
S_4	0.4	0	0.6	0			●	
S_5	0	0.8	0.2	0		●		
S_6	0.4	0	0.6	0			●	
S_7	0	0.6	0.4	0		●		
S_8	0.5	0	0.4	0.1	●			
S_9	0.4	0.6	0	0		●		
S_{10}	0	0	0.7	0.3			●	
Total	2.7	2.6	3.7	1.0	2	4	3	1

(2) Compare the fuzzy mode with the traditional mode to decide where to travel reasonably, and give the reasons;

(3) Suppose that it will cost CNY30/person, CNY15/person, CNY20/person, and CNY25/person respectively to travel to Xiachuan Island, Botanical Gardens, Maofeng Mountain and Baishui Village, use the principle of the fuzzy sample median, and choose reasonably a travel destination, and explain the reasons.

3. Shooting at the target until hitting it. Suppose the shooting is independent, the probability of hitting the target every time is p; A says the event of "shooting about 3 times and hitting the target", $A = \frac{0.2}{1} + \frac{0.7}{2} + \frac{1}{3} + \frac{0.7}{4} + \frac{0.1}{5}$, solve the probability of A.

4. Given that the reject rate of a certain product is 0.01, take randomly 100 from it to test, A represents the event of "about 2 waste products among the samples", setting $A = \frac{0.6}{1} + \frac{1}{2} + \frac{0.5}{3}$. Solve the probability of A.

5. Equipping 5 water supply facilities of the same type in a building, the surveys show that at any time t, the probability of each device being used is 0.1. A represents the event of "About 2 devices are being used", let $A = \frac{0.3}{1} + \frac{1}{2} + \frac{0.2}{3}$. Solve the probability of A.

6. Among the rock specimens taken from an area, 80% of them are mineralized, and 20% of them are not mineralized. Take 10 specimens randomly from them, A represents the event of "all the taken rock specimens are nearly mineralized", and B represents the event of "about 6 of the taken rock specimens are mineralized". Set the taken mineralized rock specimens as domain $U = \{0, 1, 2, 3, 4, \ldots, 10\}$, and Let

$$A = \frac{0.2}{5} + \frac{0.4}{6} + \frac{0.6}{7} + \frac{0.8}{8} + \frac{0.9}{9} + \frac{1}{10},$$

$$B = \frac{0.5}{4} + \frac{0.5}{5} + \frac{1}{6} + \frac{0.8}{7} + \frac{0.5}{8}.$$

Solve the probability of A, B.

7. There are 20 balls with the same texture and size, but with different colors in a box, 12 of which are red balls, and the rest are blue. Now take 5 balls from the box, what is the probability with about 3 red balls? Use the fuzzy statistics method to determine the corresponding fuzzy number for the event of "about 3 red balls among 5 taken balls" to solve the above problems.

Reading and Reflecting

An Application of Fuzzy Probability Statistics Methods in the Measurement of Penalty

The fuzzy probability statistics method is trying to be applied to the judicial practice of sentencing, i.e., by using fuzzy probability statistical theory and methods, a mathematical model will be established for intentional manslaughter sentencing, and the case will be analyzed.

First of all, a fuzzy set of an objective function is established elementarily to target the sentencing outcome, and the corresponding variables and their implications are determined in the model; Secondly, determine each variable's data value and its weight of inherent law relationship with the objective function; and, finally, establish a mathematical model by analyzing and inspecting practical cases, and modifying to make it perfect.

The following cases of an intentional manslaughter crime are some concrete applications by fuzzy probability statistics theory in the sentencing.

(1) **The Sentencing Circumstances' System for an Intentional Manslaughter Crime**

   1) The criminal motive and purpose: It mainly consists of revenge, bullying, and outrage manslaughter;
   2) The criminal age: There are mainly minors, elders over the age of 65, and adults;
   3) The criminal means: There are mainly violence, coercion, deception and lure;
   4) The criminal locations: It mainly consists of in public, public and hidden places;
   5) The criminal damage consequences: There are mainly attempted murders, one murderee, two murderees, more than two murderees;
   6) The criminal victims: There are mainly acquaintances, general awareness, strangers;
   7) The attitudes after a crime: They are mainly divided into confession, denial, and framing others.

(2) **The Membership Degree to Determine the Specific Sentencing Circumstances**
   Confirmation of a membership degree is based on the investigation into general population as the object. The fuzzy statistics method will determine its effect on the case.

   In Tables 4.9, 4.10, 4.11, 4.12, 4.13, 4.14 and 4.15, the data are obtained through surveys of 150 people, which would gain the statistical probability of the bad social impact degree that people would regard. Then with the statistical probability based on the actual weights of the experiences, solve the membership degree of bad social impacts. Specifically among 150 people, 138 people would think the criminal motive is "particularly bad" for "vicious

Table 4.9 The criminal motive's membership degrees to bad social impacts

	Particularly bad	Worse	Generally bad	Membership degree
Vicious revenge	0.92	0.07	0.01	0.81
Bullying	0.93	0.06	0.01	0.82
Outrage manslaughter	0.15	0.24	0.61	0.18

Table 4.10 The criminal age's membership degrees to bad social impacts

	Particularly bad	Worse	Generally bad	Membership degree
Minors	0.19	0.27	0.54	0.21
Elders of over 65	0.23	0.32	0.45	0.25
Adults	0.76	0.13	0.11	0.68

Table 4.11 The criminal means' membership degrees to bad social impacts

	Particularly bad	Worse	Generally bad	Membership degree
Violence	0.64	0.22	0.14	0.581
Coercion	0.52	0.24	0.24	0.49
Deception and lure	0.33	0.29	0.38	0.33

Table 4.12 The criminal location's membership degrees to bad social impacts

	Particularly bad	Worse	Generally bad	Membership degree
In public	0.87	0.11	0.02	0.77
Public	0.61	0.32	0.07	0.56
Hidden places	0.31	0.33	0.36	0.31

Table 4.13 The criminal damage consequence's membership degrees to bad social impacts

	Particularly bad	Worse	Generally bad	Membership degree
Over two murderees	0.95	0.04	0.01	0.84
Two murderees	0.93	0.06	0.03	0.82
One murderee	0.90	0.07	0.03	0.79
Attempted murders	0.50	0.26	0.24	0.4

Table 4.14 The criminal victims

	Particularly bad	Worse	Generally bad	Membership degree
Acquaintances	0.97	0.02	0.01	0.857
General awareness	0.87	0.11	0.02	0.77
Strangers	0.83	0.13	0.04	0.741

Table 4.15 The attitudes after a crime

	Particularly bad	Worse	Generally bad	Membership degree
Confession	0.16	0.18	0.66	0.18
Denial	0.74	0.24	0.02	0.67
Framing others	0.92	0.06	0.02	0.81

revenge". 11 people consider it "worse". 1 person think it "generally bad". So the statistical probabilities are obtained as follows: 0.92, 0.07 and 0.01. According to the different degrees of bad social impacts, weighted respectively by: 21, 2, 1, the weighted values multiplied by the probabilities, are divided by the weighted value of 24, you can come to the membership degrees.

(3) **Determine the Sentencing Circumstances' Weights in Overall Social Harmfulness**

The sentencing circumstances' weights are based on past experiences. They are presented with a certain degree of "subjective" factors, but this "subjective" should be set up on the basis of a large number of past cases and legal concepts with in-depth analyses and researches, which is a reflection of objective realities, as shown in Table 4.16.

Table 4.16 The main sentencing weights on intentional manslaughter crime in overall social impacts

Sentencing	Motive	Age	Means	Location	Consequence	Victim	Attitude
Weight	0.16	0.04	0.26	0.11	0.24	0.14	0.05

(4) **Analyses of a Concrete Case**

(1) A Case in Brief

Ms. Xiuzhen Li and Mr. Jie Guo, the defendants, were villagers in Beibaoxiang, Qingshuihe County. They both had kept an improper relation between men and women for many years, which led to the emotive disharmony between Ms Xiu-zhen Li and her husband, Mr. Li, often fighting. On March 14, 2007, Ms Xiu-zhen Li had a quarrel with Mr. Li about his borrowing money etc. for gambling. The next afternoon, Ms Xiuzhen Li found and asked Mr. Jie Guo to help "punish" her husband. At midnight that night, while Mr. Li was sleeping at home, she called in Mr. Jie Guo armed with a stick in his hand. While Ms Xiu-zhen Li was shining her flashlight over Mr. Li, Mr. Jie Guo beat Mr. Li on the head with his stick. Mr. Li was swaying in bed when he was being hit on the Kang. Ms. Xiu-zhen Li and Mr. Jie Guo continued to beat Mr. Li on the head, he fell to the ground. Ms. Xiu-zhen Li found him still breathing. She found a nylon rope from home, and together with Mr. Jie Guo's help, she strangled her husband to death. Afterwards, the accused confessed to the crime. On December 5, The Hohhot Intermediate People's Court made a first-instance judgment according to law: Ms. Xiu-zhen Li,

Table 4.17 The main sentencing weights on intentional manslaughter crime in overall social impacts

Attitude	Motive	Age	Means	Location	Consequence	Victim
Confession	Vicious revenge	Adults	Violence	Hidden places	One murderee	Acquaintance

the defendant, is guilty of a voluntary manslaughter crime, sentenced to death, and deprived of political rights for life; Mr. Jie Guo, the defendant, is guilty of an intentional manslaughter crime, sentenced to death, suspended for two years, and deprived of political rights for life. (China Court Website—Case Library)

(2) Analyses

The sentencing circumstances of the accused, as shown in Table 4.17.

The membership degrees in specific sentencing circumstances of the above case are respectively as follows:

The criminal motive: $0.81 \times 0.16 \approx 0.13$; the offender's age: $0.68 \times 0.04 \approx 0.03$; the means: $0.58 \times 0.26 \approx 0.15$; the criminal location: $0.31 \times 0.11 \approx 0.03$; the criminal damage consequences: $0.79 \times 0.24 \approx 0.19$; victim: $0.85 \times 0.14 \approx 0.12$; the attitude after the crime: $0.18 \times 0.05 \approx 0.01$.

The synthetical membership degree of the whole facts is the sum of all membership degrees of all sentencing circumstances in the case, that is

$$0.13 + 0.03 + 0.15 + 0.03 + 0.19 + 0.12 + 0.01 = 0.66.$$

Article No. 232 of the Criminal Law stipulates: "The punishment for intentional murder is death, life imprisonment or imprisonment of at least ten years; the minor shall be sentenced to imprisonment between three to ten years." According to the specific facts of the case, it is not "minor" intentional homicide, which corresponds to the statutory sentence is "the death penalty, life imprisonment or imprisonment of at least ten years".

Reference to the comprehensive membership degree (0.66) to the case's facts, combined with the intentional homicide in this case corresponding to the penalty (death penalty, life imprisonment or imprisonment of at least ten years), the defendants shall be declared such penalty in this case.

Taking into account the severity factor, the comprehensive membership degree (0.66) is too big to social impacts of this case. The punishment shall be seriously prescribed as emphasis (the death penalty, life imprisonment or imprisonment of at least ten years), therefore, it is more appropriate to declare punishment for the main perpetrator "the death penalty" and for the accessory "life imprisonment", matched with the first-instance judgment. The results of this evaluation are relatively objective, which can be less subjective than the judgment by a judge with personal experiences.

Summary

I. The Knowledge Structures of This Chapter

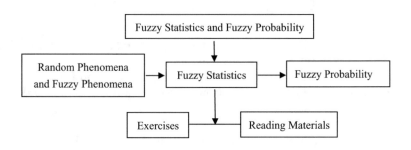

II. Review and Reflection

1. Understand the relation and difference between random and fuzzy phenomena.
2. Give an example of the fuzzy statistics methods of membership function, and review its basic steps in seeking answers.
3. Give an example for presentation and operations of fuzzy data.
4. Use the information technology (such as Excel) to process or render the descriptive statistics of fuzzy data.
5. Give an example for basic methods to solve fuzzy events' probability.

Chapter 5
Fuzzy Linear Programming

5.1 Ordinary Linear Programming and Fuzzy Linear Programming

Linear Programming (LP for short) is the ability by which the human being is endowed to represent general goals, in the face of immense complexity of practice, to achieve the goals best and give a detailed decision making solution. This ability began in 1947 shortly after World War II. With the rapid development of computer skills, due to the too rapid development of decision science, it cannot be estimated that LP and its expansion model has created a lot of wealth for mankind. We should bear in mind the work of the pathfinders: John von Neumann (United States of America), Leonid Vitaliyevich Kantorovich (Russia), Wassily Leontief (Russia), Tjalling C. Koopmans (United States of America), and George Bernard Dantzig (United States of America). First two and the last one are famous mathematicians. The three of the middle are Nobel Prize winners.

LP has been described in Mathematics 5 of the high school in China curriculum standard experiment textbook in which the objective function and the constraints are identified. Their form is:

$$\text{Objective: max } z = c_1 x_1 + c_2 x_2$$
$$\text{Constraint 1: } a_{11} x_1 + a_{12} x_2 \leq (\text{or} \geq) b_1,$$
$$\text{Constraint 2: } a_{21} x_1 + a_{22} x_2 \leq (\text{or} \geq) b_2.$$

LP is a deterministic optimization problem. But complex real-world optimization problems are often uncertain. And this uncertainty performs in practical applications for which the linear constraints and the linear objective function are fuzzy, which creates a fuzzy linear optimization problem. Such situations should be dealt with by the method of fuzzy sets.

What this section introduces is that the linear objective function and the linear constraint function should allow themselves a certain range of fuzzy optimization,

© Springer International Publishing AG, part of Springer Nature 2018
H.-R. Lin et al., *Fuzzy Sets Theory Preliminary*,
https://doi.org/10.1007/978-3-319-70749-5_5

which we call it fuzzy linear programming (recorded as FLP). It is by introducing a membership function to educe a new clear LP. This clear optimal solution for the LP is equivalent to the fuzzy optimal solution for the original FLP.

5.2   Binary FLP and Graphics

In a number of practical problems, the mathematical model of FLP is:

$$(\text{FLP}) \quad \text{Objective: } \widetilde{\max} \ z = c_1 x_1 + c_2 x_2$$
$$\text{Constraint 1: } a_{11} x_1 + a_{12} x_2 \widetilde{\leq}(\text{or} \widetilde{\geq}) b_1,$$
$$\text{Constraint 2: } a_{21} x_1 + a_{22} x_2 \widetilde{\leq}(\text{or} \widetilde{\geq}) b_2.$$

The constraint "$\widetilde{\leq}$" or "$\widetilde{\geq}$" means "about smaller than or equal to" or "about bigger than or equal to", which is a fuzzy concept and denotes a telescopic constraint. $\widetilde{\max} \ z$ denotes solving the maximal value for Objective z. There are some fuzzy unequal relation in this model. What methods should we use to describe them?

Supppose that $d = d_i$ is the maximal telescopic value at the right end coefficient b_i of the constraint function, this value is set by people according to actual situation. If the value of the constraint function exceeds the maximal acceptable range, it is regarded as an "unqualified" value, denoted by the membership degree: "0"; if it falls completely within the constraint range, it is judged as a "qualified" value, denoted by the membership degree: "1"; if the value is between the "qualified" and the "unqualified", it can be denoted by a numerical value in the interval [0, 1] to reflect its qualified degree.

In this way, the linear membership function $\widetilde{D}_i(x)$ is introduced as follows:

$$\widetilde{D}_i(x) = f_i(ax)$$

$$= f_i(y) = \begin{cases} 1, & a_i x \leq b_i, \\ 1 - \dfrac{(a_{i1} x_1 + a_{i2} x_2 - b_i)}{d_i}, & b_i < a_i x \leq b_i + d_i, \\ 0, & a_i x > b_i + d_i (i = 1, 2), \end{cases} \quad (5.1)$$

its image is shown in Fig. 5.1.

The flexibility of the constraint function is inevitably sure to lead to flexibility of the objective function. Suppose that z_0 and z_0' respectively are the basic value of the

Fig. 5.1 The relation between $\widetilde{D}_i$ and y_i

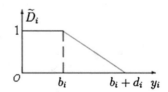

constraint function and the maximal telescopic value of it, and select $d_0 = z_0' - z_0$ as the maximal telescopic value of the objective function. Then, if the objective function's maximal value is smaller than z_0, it is identified as an "unsatisfactory" value, denoted by "0"; if it is bigger than z_0, it is judged to be a "satisfactory" value, denoted by "1"; if the value is between the "unsatisfactory" and the "satisfactory", it can be denoted by a numerical value in the interval [0, 1] to reflect its degree of satisfaction. Then it introduces another linear membership function $\widetilde{G}(x)$:

$$\widetilde{G}(x) = g(cx)$$

$$= g(z) = \begin{cases} 0, & cx \leq z_0, \\ \dfrac{(cx - z_0)}{d_0}, & z_0 < cx \leq z_0 + d_0, \\ 1, & cx > z_0 + d_0, \end{cases} \tag{5.2}$$

its image is shown in Fig. 5.2.

Fig. 5.2 The relation between $\widetilde{G}_i$ and z

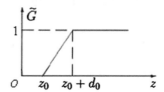

We find that: When $\widetilde{G}(x) = 1$, $\widetilde{D}(x) = 0$. In order to make the target value bigger than z_0, $\widetilde{D}(x)$ must be reduced. Here, $\widetilde{D}(x)$ denotes the "qualified" degree to the constraints. $\widetilde{D}(x)$ denotes the "satisfactory" degree to the objective. A natural idea is to make the fuzzy constraints and the fuzzy objective as maximal as possible. So, a fuzzy judgement of FLP is defined as: $\widetilde{D}_f = \widetilde{D} \cap \widetilde{G}$. Suppose that the optimal solution of FLP is x^*, then

$$\begin{aligned} \widetilde{D}_f(x^*) &= \max_{x \geq 0} \widetilde{D}_f(x) = \max_{x \geq 0} \{\widetilde{G}(x) \wedge \widetilde{D}(x)\} \\ &= \max_{x \geq 0} \{\widetilde{G}(x) \wedge \widetilde{D}_1(x) \wedge \widetilde{D}_2(x)\}. \end{aligned} \tag{5.3}$$

Can it be converted into a general LP problem then if we let $\lambda = \widetilde{G}(x) \wedge \widetilde{D}_1(x) \wedge \widetilde{D}_2(x)$? The answer is yes, interested readers can try it on their own.

In fact, the (FLP) solution is to have the constraint flexibility index corresponded to the fuzzy constraint; accordingly, the objective flexibility index is selected to correspond to the fuzzy objective. Their membership functions are defined respectively. An FLP problem can be converted into an LP problem to solve. The calculating program is complex, which needs to be done by computer with LP algorithmic program. Inspired by an LP graphical solution, it is given below for a two-dimension FLP problem. The FLP problem's graphical solution has the same advantage as that of the LP problem's graphical solution, simple and intuitive. It is no need to use any computer algorithm programs to solve problems. It can make graphical intuitive analyses of optimization problems.

The FLP graphical solution is explained with the following examples.

Suppose that the FLP problem is shown as follows:

Example 5.1 Solve:

$$\text{Objective:} \quad \max z = 3x_1 + 4x_2$$
$$\text{Constraint 1:} \quad x_1 + 4x_2 \lesssim 160,$$
$$\text{Constraint 2:} \quad x_1 \lesssim 40,$$
$$x_1 \geq 0, x_2 \geq 0.$$

Two flexibility indices d_1 and d_2 of the constraints are respectively 40 and 5. The FLP graphical solution is shown in Fig. 5.3. The solution steps and ideas are explained as follows:

1. Use the LP graphical solution to solve the basic values of the constraints, i.e., solve:

$$\text{Objective:} \quad \max z = 3x_1 + 4x_2$$
$$\text{Constraint 1:} \quad x_1 + 4x_2 \leq 160,$$
$$\text{Constraint 2:} \quad x_1 \leq 40,$$
$$x_1 \geq 0, x_2 \geq 0.$$

As shown in the solid line part of Fig. 5.3, the feasible region is a convex quadrilateral ABCD. The contour line IH of the objective function intersects the feasible region at the vertex B, which is the optimal solution for $x_1 = 40$, $x_2 = 30$, the optimum value $z_0 = 240$.

Fig. 5.3 An example by the FLP graphical solution

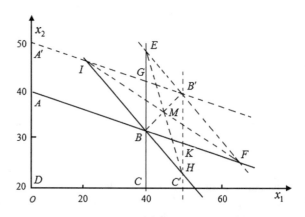

2. Use the LP graphical solution to solve the telescopic values when the constraints extend to the maximal range, i.e., solve:

$$\text{Objective:} \quad \max z = 3x_1 + 4x_2$$
$$\text{Constraint 1:} \quad x_1 + 4x_2 \leq 160 + 40,$$
$$\text{Constraint 2:} \quad x_1 \leq 40 + 5,$$
$$x_1 \geq 0, x_2 \geq 0.$$

As shown in the dashed line part of Fig. 5.3, the feasible region has expanded into the quadrilateral A′B′C′D. The isoline EF of the objective function intersects the feasible region at the vertex B′, which is the optimal solution for $x_1 = 45$, $x_2 = 38.75$, the optimum value $z_0 = 290$.

Take $d_0 = z'_0 - z_0 = 50$ as the telescopic index of the objective function z.

3. When the constraints' telescopic ranges are increased, the optimal solution moves from B to B′. When it moves, it is leaving from the basic constraints 1 and 2. According to Eq. (5.1) and Fig. 5.1, $\widetilde{D}_1(x)$ and $\widetilde{D}_2(x)$ are decreasing while it is closer to the maximal telescopic objective line EF. According to Eq. (5.2) and Fig. 5.2, $\widetilde{G}(x)$ is increasing. The problem now is to solve a suitable point to satisfy Eq. (5.3), that is, to solve the optimal solution x^*. For this, the following definition is given.

Definition 5.1 The telescopic range of each constraint and objective is defined as the domain between the two parallel isolines, called the telescopic domain; the intersections of two different telescopic domains make a parallelogram, called the telescopic parallelogram.

Obviously, the fuzzy constraint's or the fuzzy objective's membership degrees are 0 and 1 respectively for points on the two sidelines of the telescopic domain. They correspond to the points: $b_i + d_i$ and b_i on the abscissa of Fig. 5.1 and the points: z_0 and $z_0 + d_0$ on the abscissa of Fig. 5.2 respectively. The distance between two parallel lines corresponds to the telescopic index.

Definition 5.2 In Fig. 5.3, the parallelograms of BGB′K, BEB′H, and BIB′F are telescopic parallelograms.

In the FLP Graphical Solution, three sidelines of the telescopic domains with 0 membership degree form a triangle. One of its vertices is the LP problem's optimal solution when the fuzzy constraint extends to the maximal rang. The triangle is called a telescopic triangle.

In Fig. 5.3, △IB′H is a telescopic triangle. The FLP problem's solution is that $\widetilde{D}_f(x)$ is increasing gradually from 0 to the optimal solution with a maximal value, and then it is decreasing to 0 in the moving progress when the point (B) on one side of the triangle moves to the opposite vertex (B′). The Optimal solution is within the range of the telescopic triangle.

Definition 5.3 Each angle dividing line of the telescopic triangle is called the equal membership angle-dividing line. In practice, we can find out the telescopic parallelogram containing the angle, and connect the diagonal containing the angle vertex.

The FLP Graphical Solutions of optimal solutions are explained as follows:

Draw the equal membership angle-dividing line for an angle of the telescopic triangle. In Fig. 5.3, the telescopic parallelogram BIB'F contains $\angle$B'IH. Connect the diagonal line IF, as shown by the dashed in Fig. 5.3. It is the same with another solution of $\angle$IHB', which is contained in the corresponding parallelogram BEB'H. Connect the diagonal line EH, which intersects the line IF at M. M is the optimal solution x^*, which satisfies the Eq. (5.3). Actually the three equal membership angle-dividing lines of the telescopic triangle intersect at a single point. The intersection can be found by any of the two lines. In Fig. 5.3, the point M's coordinates are (42.5, 34), so $z^* = cx^* = 263.5$.

In sum, the steps for the FLP's graphical solutions are summarized as follows:

Step 1. Use the classic LP graphical solutions to solve the LP's graphical solution with basic constraints (for solutions, see Sect. 3.3.2 of *Mathematics 5* of the standard experiment textbook for senior middle school curriculum).

Step 2. Use the classic LP graphical solutions to solve the LP's graphical solution when the constraints extend to a maximal range.

Step 3. Draw two equal membership angle-dividing lines of the telescopic triangle. The intersection is the optimal solution, by which the optimal value (maximum or minimum) of the objective function can be found.

5.3   Application and Excel Solutions of Binary FLP

Example 5.2 Vegetarian Issues: If the independent choices of vegetables and fruit juice are restricted, a vegetarian can buy x units of vegetables and y units of fruit juice once a week. According to the required consumption of the vitamins' specific gram equivalent, the vegetarian food choice should be subject to further restrictions, which are often fuzzy. In this way, he will choose a little bit more than 40 g equivalents of vitamin A (set the telescopic index for $d_A = 0.2$), about a little bit more than 50 g equivalents of vitamin B ($d_B = 0.5$), about a little bit more than 70 g equivalents of vitamin C ($d_C = 0.5$), a little bit more than 10 g equivalents of vitamin D ($d_D = 0.2$), and a little bit more than 60 g equivalents of vitamin E ($d_E = 0.2$) food. Every unit of vegetables contains respectively 0.1, 0.2, 0.04, 0.1 and 0.06 g equivalents of vitamin A, B, C, D and E. Every unit of fruit juice contains respectively about 0.05, 0.15, 0.2, 0.1 and 0.1 g equivalents of vitamin A, B, C, D and E. The price of 1 unit of vegetables is 2 cents, and the price of 1 unit of fruit juice is 3 cents. So the vegetarian issues are the vegetarians' purchase quantity of vegetables and fruit juice that can meet their body needs for vitamins and cost as little as possible (set the telescopic index for $d_z = 0.5$).

1. **Analyses**: The given data are listed in the table below

Any positions	A	B	C	D	E	Price $
Vegetables	0.1	0.2	0.04	0.1	0.06	2
Fruit juice	0.05	0.15	0.2	0.1	0.1	3
Human needed vitamins	$40 - d_A$	$50 - d_B$	$50 - d_B$	$10 - d_D$	$60 - d_E$	

2. **Establishment of Fuzzy Linear Programming Model**

Suppose that a vegetarian can eat x units of vegetables and y units of fruit juice every day, thus, the vegetarian has to buy x units of vegetables and y units of fruit juice every day. The money that he pays for them is the objective function $z = 0.02x + 0.03y$.

Our problem is to solve x and y, and make $z = 0.02x + 0.03y$ (objective function) with the fuzzy minimum (i.e., $\underset{min}{Z\sim}$), which can satisfy the constraints:

$$
\begin{aligned}
A \quad & 0.1x + 0.05y \gtrsim 40 \\
B \quad & 0.2x + 0.15y \gtrsim 50 \\
C \quad & 0.04x + 0.2y \gtrsim 70 \\
D \quad & 0.1x + 0.1y \gtrsim 10 \\
E \quad & 0.06x + 0.1y \gtrsim 60 \\
& x \geq 0, y \geq 0.
\end{aligned}
\tag{5.4}
$$

It sums up as follows:

$$
\begin{aligned}
\underset{min}{Z\sim} &= 0.02x + 0.03y \\
\text{s.t.} \quad & 0.1x + 0.05y \gtrsim 40, \\
& 0.2x + 0.15y \gtrsim 50, \\
& 0.04x + 0.2y \gtrsim 70, \\
& 0.1x + 0.1y \gtrsim 10, \\
& 0.06x + 0.1y \gtrsim 60, \\
& x \geq 0, y \geq 0.
\end{aligned}
\tag{5.5}
$$

3. Using Excel to Solve FLP Problems

(1) Solving the Classical LP

For reference to the senior middle school textbook in China *Mathematics 5*, using Excel to solve:

$$\begin{aligned}
\min \quad & z = 0.02x + 0.03y \\
\text{s.t.} \quad & 0.1x + 0.05y \geq 40, \quad 0.2x + 0.15y \geq 50, \\
& 0.04x + 0.2y \geq 70, \quad 0.1x + 0.1y \geq 10, \\
& 0.06x + 0.1y \geq 60, \quad x \geq 0, y \geq 0,
\end{aligned} \tag{5.6}$$

solve and find:

	B12		f_x	=0.02*x+0.03*y	
	A	B		C	D
1	variables	values			
2	x	142.8571429			
3	y	514.2857143			
4					
5	subject to				
6		40		40	
7		105.7142857		50	
8		108.5714286		70	
9		65.71428571		10	
10		60		60	
11					
12	object value (min)	18.28571429			
13					

The optimal solution is x = 142.857142429, y = 514.2857143, the minimal value of the objective function is 18.28571429.

In order to restore the original established initial value, we choose to save the results here.

(2) Solving the General LP with Added Relaxation

The right amount of Inequalities (5.6) represents the vitamins' values that the human beings need. If the telescopic index is considered respectively, namely: $40 - d_A = 39.8$, $50 - d_B = 49.5$, $70 - d_C = 69.5$, $10 - d_D = 9.8$, $60 - d_E = 59.8$. So sovle the solution below:

$$\begin{aligned}
\min \quad & z = 0.02x + 0.03y \\
\text{s.t.} \quad & 0.1x + 0.05y \geq 39.8, \quad 0.2x + 0.15y \geq 49.5, \\
& 0.04x + 0.2y \geq 69.5, \quad 0.1x + 0.1y \geq 9.8, \\
& 0.06x + 0.1y \geq 59.8, \quad x \geq 0, y \geq 0.
\end{aligned} \tag{5.7}$$

Similar to the first round of the process, just modify the values from C_6 to C_{10} in the above table as the relaxation requirements. The settings for the solver parameters are the same as those of the first round, without modification. The results are shown in the following table:

	B12		$\mathbb{Q}$ fx	=0.02*x+0.03*y
	A	B		C
1	variables	values		
2	x	141.4285714		
3	y	513.1428571		
4				
5	subject to			
6		39.8		39.8
7		105.2571429		49.5
8		108.2857143		69.5
9		65.45714286		9.8
10		59.8		59.8
11				
12	object value (min)	18.22285714		

The optimal solution is $x = 141.4285714$, $y = 513.1428571$, the minimal value of the objective function is 18.22285714.

3. Sovle the Solution after Adding Variables.
The objective change is $d_z = -0.5$. Based on the fuzzy decision formula, after introducing a variable λ, the LP is obtained with fuzzy parameters:

max $G = \lambda$

S.t. $0.1x + 0.05y - 0.2\lambda \geq 39.8,$ $0.2x + 0.15y - 0.5\lambda \geq 49.5,$

$0.04x + 0.2y - 0.5\lambda \geq 69.5,$ $0.1x + 0.1y - 0.2\lambda \geq 9.8,$ (5.8)

$0.06x + 0.1y - 0.2\lambda \geq 59.8,$ $0.02x + 0.03y + 0.0628\lambda \leq 18.2857,$

$x \geq 0, y \geq 0, 0 \leq \lambda \leq 1.$

Similar to the first round of the process, first establish the original data table below (which can be modified on the basis of the data in the second round table), and list the slope of the straight line.

	B13	▾	⊕ fx	=z		
	A		B		C	D
1	variables		values			
2	x		1			
3	y		1			
4	z		0			
5						
6	subject to		0.15		39.8	
7			0.35		49.5	
8			0.24		69.5	
9			0.2		9.8	
10			0.16		59.8	
11			0.05		18.2857	
12						
13	Z max value		0			
14	object min value		0.05			

Click "solve", we get

	B13	▾	⊕ fx	=z	
	A		B		C
1	variables		values		
2	x		142.1430209		
3	y		513.7144144		
4	z		0.500113342		
5					
6	subject to		39.80000014		39.8
7			105.2357097		49.5
8			108.178547		69.5
9			65.48572086		9.8
10			59.80000003		59.8
11			18.28569997		18.2857
12					
13	Z max value		0.500113342		
14	object min value		18.25429285		

It can be seen that the solution is $x = 142.1430209$, $y = 513.7144144$, the minimal value of the objective function is 18.25429285.

To sum up: $\lambda = 0.500113342$, $x = 142.1430209$, $y = 513.7144144$ is the optimal solution to the FLP, the value is 18.25429285.

Note: The graphical solution is used to solve Example 5.2.

1. Solving the Ordinary LP (5.6)

First draw the feasible region in Fig. 5.4 (the right angle portions of A, B, C, and D).

Change $z = 0.02x + 0.03y$ into $y = -\dfrac{0.02}{0.03}x + \dfrac{z}{0.03}$. This is a straight line with $-\dfrac{2}{3}$ as the slope and $\dfrac{z}{0.03}$ as an intercept in Axis y. When z changes, a group of

Fig. 5.4 The constraint function of programming (5.6)

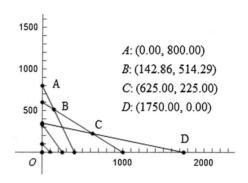

$A: (0.00, 800.00)$
$B: (142.86, 514.29)$
$C: (625.00, 225.00)$
$D: (1750.00, 0.00)$

mutual parallel isolines can be obtained. When the coordinates of the points: A, B, C, and D are substituted into the objective function $z = 0.02x + 0.03y$, the optimal solution B can be easily obtained by comparing their values. The coordinates are $x = 142.86$, $y = 514.29$, and the optimal solution $z = 18.2859$.

2. Solving the Ordinary LP (5.7) with Additional Relaxation

First the feasible region is made in Fig. 5.5 (the right-up corner portions of E, F, G, and H):

Similarly, comparing the function values of the points: E, F, G, and H, it is easy to get the optimal solution. The coordinates are $x = 141.43$, $y = 513.14$, and the optimal solution $z = 18.2228$.

3. Solving the Fuzzy Programming (5.8) with Additional Variable λ

Make two equal membership angle-dividing lines of the telescopic triangle. Their intersection is the optimal solution, as shown in Fig. 5.6. In order to obtain a more accurate image, the following explanation is made especially.

For the "Programming (5.6)", there are only two effective constraints $\begin{cases} 0.1x + 0.05y \geq 40 \\ 0.06x + 0.1y \geq 60 \end{cases}$, which are demonstrated by the solid lines in Fig. 5.6. The intersection coordinates of B are (142.86, 514.29).

Fig. 5.5 The constraint function of programming (5.7)

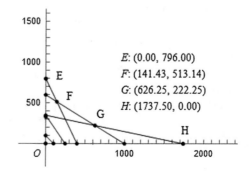

$E: (0.00, 796.00)$
$F: (141.43, 513.14)$
$G: (626.25, 222.25)$
$H: (1737.50, 0.00)$

Fig. 5.6 The FLP graphic
methods

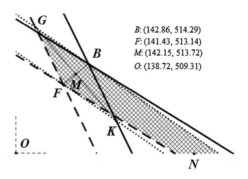

For the "Programming (5.7)", there are only two effective constraints
$\begin{cases} 0.1x + 0.05 \geq 39.2 \\ 0.06x + 0.1y \geq 59.8 \end{cases}$, which are demonstrated by the dashed lines in Fig. 5.6. The
intersection coordinates of F are (141.43, 513.14).

There are more lines in the feasible region. Here we only draw the valid constraint
lines and the isolines of the objective function, where the isolines of the objective
function are represented by the dotted lines. As the values of the elasticity indices:
$d_A = -0.2$, $d_B = d_C = -0.5$, $d_E = d_F = -0.2$ are very small, so the picture has
been magnified many times in order to be able to see clearlier the difference between
$\begin{cases} 0.1x + 0.05y = 40 \\ 0.1x + 0.05y = 39.8 \end{cases}$ and $\begin{cases} 0.06x + 0.1y = 60 \\ 0.06x + 0.1y = 59.8 \end{cases}$, as shown in Fig. 5.6.

Based on fuzzy graphic thinking, a triangle composed of three intersections of 0
membership degree lines is a equal membership triangle, as shown in the shadow of
Fig. 5.6 (there is an intersection N in the distance). When the constraint function
reduces, the feasible region expands. The optimal point moves from B to F gradually while its membership degree goes from "1" to "0". After drawing two
angle-dividing lines of the telescopic triangle, we find that their intersection M is
just the intersection of two diagonals of the parallelogram BGFK. Because of the
nature that the parallelogram diagonals divide each other equally, it decides the
point M (142.15, 513.72) is sure to bisect the line BF, so substituting the values into
the objective function to obtain the optimal value: 18.2546.

Reflection

(1) **Analyze**: If the right side of the inequality for human body needed vitamins in
Model (5.6) is changed respectively into $40 + d_A = 40.2$, $50 + d_B = 50.5$,
$70 - d_C = 69.5$, $10 + d_D = 10.2$, $60 + d_E = 60.2$. The value of the objective
function is changed into $d_Z = 0.5$.

The final result: $\lambda = 1$, $x = 142.8571$, $y = 514.2857$, the optimal value is
18.28571.

Students can use the above Graphic Methods to check.

The result is compared with the telescopic Programming $x = 144.2857$,
$y = 515.4286$, the minimal value of the objective function is 18.34857, which has a

certain difference. After an increase in elasticity constraints, the programming's optimal value is not better than before, which is very inconsistent with reality. In fact, after an increase in elasticity indices, the feasible region becomes smaller, which is nearly consistent with the programming results without retractility: $x = 142.8571$, $y = 514.2857$, the minimum value of the objective function is 18.28571. This means that the telescopic indices we get is incorrect in the pointed direction.

(2) **Comparing**: Solving the programming (5.3.2) with the Excel solution, it concludes that $x = 142.1430$, $y = 513.7144$, the optimal value is 18.2543. The results are consistent with that of the graphical methods.

Students can use the Excel solution to check.

Exercise 5

1. Assume that a fashion furniture company produces each table with profit of 7 \$, each chair with profit of 5 \$. It makes a table of raw materials with 5 B.F (Board feet, Board sq ft), and a chair with 2 B.F, and weekly all of raw materials with over 50 B.F, floating up to 5 B.F; it takes 2 h to produce a table, and 3 h to produce a chair, and it weekly can reach over 42 working hours, Floating up to 8 h. How many tables and chairs will the company produce weekly to make the maximum profits?
2. A vegetarian only eats salad dish and bread who is provided to consume daily vitamin A, B and C at a little bit more than 70, about 86 and almost 62 units respectively. A pound of salad dish contains 7 units of vitamin A, 4 units of vitamin B and 3 units of vitamin C. A loaf of bread contains 2 units of vitamin A, 4 units of vitamin B and 2 units of vitamin C. A loaf of bread costs CNY3, and a pound of salad dish costs CNY0.60. Solve the least daily diet of bread and salad dishes for the vegetarian.
3. Use the graphic method to solve Problem 5.1.
4. Use the graphic method to solve Problem 5.2.
5. Use the graphic method to solve the problem

$$C\underset{max}{\sim} = 4x + 6y$$
$$s.t. \quad x + y \underset{\sim}{\leq} 8,$$
$$3.5x + 2y \underset{\sim}{\geq} 14,$$
$$x \underset{\sim}{\leq} 4, y \underset{\sim}{\leq} 7,$$
$$x \geq 0, y \geq 0.$$

6. Solve Problem 5.5 with the Excel.

Reading and Reflecting

The Origin of LP and FLP

LP began in 1947 shortly after the end of World War II, and prior to that it was not known, but there were some individual exceptions: Mr. B. J. Fourier, a famous mathematician in 1827, and Mr. L. V. Poussin, a famous Belgian mathematician in

1911, respectively wrote articles related to LP. Although they were only about it, after 1947 their effects on LP were the same as the computers found in the Egyptian Tombs in 3000 BC had done on the development of the modern computers. Mr. L. V. Kantorovich's famous treatise on LP was ignored in 1939, due to the ideological reasons of the former Soviet Union. Until 20 years later, the LP aroused people's attention again for its important progress yielded in the Western world. Mr. F. Hitchcock wrote an excellent paper on transport which was ignored in 1941. Until the late 1940s and the early 1950s, the LP aroused people's attention again because others had independently rediscovered the characteristics of transport issues. After 1947, the LP was born, which had been improved by Mr. J. v. Neumann, Mr. L. V. Kantorovich, Mr. W. Leontief, Mr. T. C. Koopmans and Mr. G. B. Dantzig. The FLP started later. In 1970, R. E. Bellman and L. A. Zadeh have first researched the optimization problem in fuzzy environments. Subsequently the FLP has been established officially by S. A. Orlovsky, C. V. Negoita, H. J. Zimmermann, H. Tanaka, D. Dubois and H. Prade etc. which has been developed in theory and methods and has a tempting prospect in application.

Summary

I. **The Knowledge Structures of This Chapter**

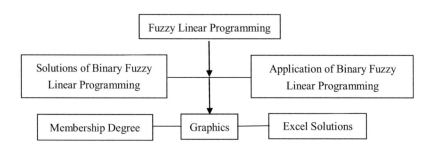

II. **Review and Reflection**
1. Understand the relation of LP and FLP.
2. Cite an example of binary LP and review its procedures for solutions.
3. Write the membership function's formula for the objective and constraint functions respectively.
4. List the steps of graphical solutions of binary FLP.
5. Solve a dual FLP problem with Excel solutions.

References

1. Zadeh, L.A.: Fuzzy Sets, Language Variables and Fuzzy Logic. Science Press, Beijing (1982)
2. Liu, Y.M., Ren, P.: The Other Half of Fuzziness—Accuracy (Academician Science Book). Qinghua University Press, Beijing (2000)
3. Gao, Q.S.: New Base of Fuzzy Set Theory: The Shortcomings and Mistakes, and Overcoming of Zadeh's Fuzzy Set Theory. Mechanical Industry Press (2006)
4. Wang, P.Z.: Fuzzy Mathematics and Optimization. Beijing Normal University Press, Beijing (2013)
5. Dou, Z.Z.: Fuzzy Logic Control Technology and Its Application. Beijing Aeronautics and Astronautics University Press, Beijing (1995)
6. Hu, B.Q.: Basic Tutorial of Fuzzy Mathematics Theory. Wuhan University Press, Wuhan (2010)
7. Juro, Terano: Fuzzy Engineering: New Century Thought Methods. Liaoning Press, Liaoning (1995)
8. Buhler, A.: Brain-like—From Science Fiction to Reality. Hunan Science and Technology Press, Hunan (2001)
9. Castilla, J.: The Cambridge Quintet—Can a Machine Think. Shanghai Science and Technology Press, China (2001)
10. Chen, Y.Z.: The Brain's Secret Black Box. Beijing Children's Press, Beijing Education Press, China (2002)
11. Wang, L.X.: Fuzzy Systems and Fuzzy Control. Qinghua University Press, Beijing (2003)
12. Yu, Y.Q.: Fuzzy Control Technique and Fuzzy Household Appliances. Beijing Aeronautics and Astronautics University Press, Beijing (2000)
13. Lin, H.R.: Thinking of Modern Mathematics Into Research-oriented Course in Middle School, Vol. 1, pp. 50–52. People's Education Press, China (2002)
14. Lin, H.R.: Talk about high school electives "fuzzy mathematics preliminary". Math. Bull. **2**, 27–28 (2002)
15. Lin, H.R.: The necessity and possibility (I). Math. Bull. **4**, 14–17 (2003)
16. Lin, H.R.: The necessity and possibility (II). Math. Bull. **5**, 10–12 (2003)
17. Lin, H.R.: Fuzzy mathematics in middle schools in China. Zhejiang Educ. Sci. **1**, 33–35 (2006)
18. Lin, H.R.: Alternative thinking on innovative education. Guangdong Educ. **9**, 54–56 (2006)
19. Wang, P.Z.: Fuzzy Set Theory and Application. Shanghai Science and technology Press, China (1983)
20. Li, H.X., Wang, P.Z.: Fuzzy Mathematics. National Defense University Press, USA (1994)
21. Wang, Z.Y., Wu, B.L.: Fuzzy Data Statistics. Harbin Industrial University Press, China (2008)
22. Zhang, Y., Zou, S.P., Su, F.: Fuzzy Mathematics Methods and Application. Coal Industry Press, China (1992)

23. Dong, R.Z., Zhang, X.L.: Study on Application of fuzzy mathematics methods in sentencing. J Henan Educ. Inst. **17**(2), 40–42 (2008). (Natural Science Edition)

24. Ministry of Education of the People's Republic of China: High School "Mathematics Curriculum Standards" (Experimental), pp. 5–10. People's Education Press, Beijing (2003)

25. Lin, H.R.: Practice of fuzzy mathematics into the high school curriculum. J. Hainan Normal Univ. **4**, 496–497 (2008). (Natural Science Edition)

26. Yang, L.B., Gao, Y.Y.: Fuzzy Mathematics—Principles and Applications, pp. 1–2. South China Polytechnic University Press, Guangzhou (1995)

27. Cao, B.Y.: Applied Fuzzy Mathematics and Systems, pp. 34–81. Science Press, Beijing (2005)

28. Lin, H.R.: Fuzzy Mathematics Preliminary—Senior Middle School Further Studied Mathematics Curriculum, pp. 1–129. Shanghai China Middle School (1997)

29. Zhang, Z.K.: Graphical methods of fuzzy linear programming. J. Qinghua Univ. **37**(9), 6–9 (1997). (Natural Science Edition)

30. Lav, B., Weiss, H.J.: Introduction to Mathematical Programming. Elsevier North Holland Inc., Amsterdam (1982)

31. Lou, S.B., Sun, Z., Chen, H.C.: Fuzzy Mathematics. Science Press, Beijing (1983)

Printed in the United States
By Bookmasters